新一代信息技术"十三五"系列规划教材

微信小程序开发
项目教程
慕课版

刘刚 | 著

人民邮电出版社

北京

图书在版编目（CIP）数据

微信小程序开发项目教程：慕课版 / 刘刚著. --北京：人民邮电出版社，2021.8（2023.12重印）
新一代信息技术"十三五"系列规划教材
ISBN 978-7-115-55065-1

Ⅰ．①微… Ⅱ．①刘… Ⅲ．①移动终端－应用程序－程序设计－教材 Ⅳ．①TN929.53

中国版本图书馆CIP数据核字(2020)第201320号

内 容 提 要

本书以一个典型项目的实现过程为主线，详细讲解了微信小程序开发技术，包括微信小程序概述，莫凡商城小程序项目任务，莫凡商城小程序的项目结构，莫凡商城首页静态布局设计，莫凡商城首页动态绑定设计，莫凡商城的注册、登录功能，莫凡商城商品详情页设计，莫凡商城获取收货地址功能设计，莫凡商城支付功能及订单详情页设计，小程序扩展应用。本书采用图、表与详细说明的示例代码相结合的叙述方式，将微信小程序设计的基本原理和知识融入项目开发实战之中，简单易懂，带着读者边做边学，快速掌握微信小程序的设计和实现，帮助读者掌握典型功能的开发，便于读者举一反三。

本书可作为对微信小程序开发有兴趣读者的自学用书，也可作为高等院校、培训机构关于微信小程序开发课程的教材。

◆ 著　　刘　刚
　　责任编辑　桑　珊
　　责任印制　彭志环

◆ 人民邮电出版社出版发行　北京市丰台区成寿寺路 11 号
　　邮编　100164　电子邮件　315@ptpress.com.cn
　　网址　https://www.ptpress.com.cn
　　天津千鹤文化传播有限公司印刷

◆ 开本：787×1092　1/16
　　印张：19　　　　　　　　　　2021 年 8 月第 1 版
　　字数：549 千字　　　　　　　2023 年 12 月天津第 6 次印刷

定价：69.80 元

读者服务热线：(010)81055256　印装质量热线：(010)81055316
反盗版热线：(010)81055315
广告经营许可证：京东市监广登字 20170147 号

前言 PREFACE

本书全面贯彻党的二十大精神,以社会主义核心价值观为引领,传承中华优秀传统文化,坚定文化自信,使内容更好体现时代性、把握规律性、富于创造性。

为什么要学微信小程序

微信小程序是微信团队在 2017 年 1 月 9 日正式发布的功能。它可以实现 App 软件的原生交互操作效果,但是不像 App 软件需要下载安装才能使用。微信小程序只需要用户扫一扫或者搜一下就可以使用,不仅符合用户的使用习惯,还解放了用户手机的内存空间,同时给企业提供了宣传自己产品的渠道。企业创建微信小程序后,其产品就可以被更多用户找到,从而宣传自己的产品。微信小程序的快速发展,为我们提供了很多就业机会。让我们赶快成为一名小程序员吧!

本书学习路径

莫凡商城小程序是贯穿本书的项目实例。读者通过学习本书内容可以完整地开发一个企业级的小程序,并在开发过程中了解和学习微信小程序的基础知识及典型模块的开发方法。本书的学习路径如下。

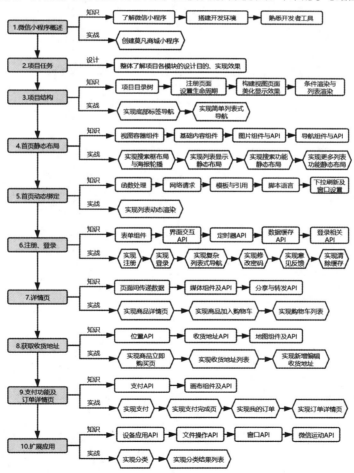

小刚（刘刚）老师简介

- 一线项目研发、设计、管理工程师，高级项目管理师、项目监理师、技术经理，负责纪检监察廉政监督监管平台、政务办公平台、政务大数据、OTO到家项目等多个大型项目的设计与开发。
- 畅销书《微信小程序开发图解案例教程（附精讲视频）》《小程序实战视频课：微信小程序开发全案精讲》《Axure RP8 原型设计图解微课视频教程（Web+App）》的作者。

平台支撑，免费赠送资源

- 全部案例源代码、全书电子教案可登录人邮教育社区（www.ryjiaoyu.com.cn）下载。
- 全书高清精讲视频课程（扫书中二维码或登录人邮学院观看，人邮学院登录方法见本书封底）。
- 问题答疑QQ群：166840379，及时解答学习过程中碰到的问题。

著者

2022年12月

目录 CONTENTS

第1章
微信小程序概述 1
- 1.1 微信小程序介绍 1
  - 1.1.1 初识微信小程序 1
  - 1.1.2 微信小程序的功能 2
  - 1.1.3 微信小程序的使用场景 3
  - 1.1.4 微信小程序的发展历程 3
  - 1.1.5 微信小程序带来的机会 4
- 1.2 微信小程序环境搭建 4
  - 1.2.1 小程序环境搭建 4
  - 1.2.2 基础技术准备 6
- 1.3 微信小程序开发者工具的使用 ... 6
  - 1.3.1 如何创建项目 6
  - 1.3.2 微信开发者工具界面 8
  - 1.3.3 常用快捷键 16
- 1.4 项目实战：创建莫凡商城小程序 16
- 1.5 小结 18

第2章
莫凡商城小程序项目任务 19
- 2.1 "我的"模块功能介绍 19
  - 2.1.1 任务1——实现底部标签导航功能 19
  - 2.1.2 任务2——实现注册功能 20
  - 2.1.3 任务3——实现登录功能 20
  - 2.1.4 任务4——实现"我的"界面列表式导航功能 20
  - 2.1.5 任务5——实现修改密码功能 21
  - 2.1.6 任务6——实现意见反馈功能 21
  - 2.1.7 任务7——实现清除缓存功能 22
  - 2.1.8 任务8——实现我的订单功能 22
- 2.2 "首页"模块功能介绍 22
  - 2.2.1 任务9——实现搜索区域布局与海报轮播功能 22
  - 2.2.2 任务10——实现图书列表显示功能静态布局与动态渲染 23
  - 2.2.3 任务11——实现图书搜索功能 23
  - 2.2.4 任务12——实现图书更多列表显示功能 23
- 2.3 "购买商品"模块功能介绍 24
  - 2.3.1 任务13——实现商品详情页功能 24
  - 2.3.2 任务14——实现商品加入购物车功能 25
  - 2.3.3 任务15——实现购物车列表功能 25
  - 2.3.4 任务16——实现商品立即购买页功能 25
  - 2.3.5 任务17——实现收货地址列表功能 25
  - 2.3.6 任务18——实现新增和编辑地址功能 26
  - 2.3.7 任务19——实现支付功能......26
  - 2.3.8 任务20——实现支付完成页功能 27

2.3.9 任务21——实现订单详情页
功能 27
2.4 "图书分类"模块功能介绍 28
2.4.1 任务22——实现图书分类
功能 28
2.4.2 任务23——实现图书分类结果
列表功能 29
2.5 小结 29

第3章
莫凡商城小程序的项目
结构 30
3.1 项目目录树结构介绍 30
3.1.1 框架全局文件 30
3.1.2 项目实战：任务1——实现
底部标签导航功能 36
3.1.3 工具类文件 38
3.1.4 框架页面文件 38
3.2 微信小程序逻辑层框架接口 39
3.2.1 使用App()函数注册小程序 ... 39
3.2.2 使用Page()函数注册页面 ... 40
3.3 微信小程序WXML视图层 42
3.3.1 WXML标签语言 42
3.3.2 动态绑定数据 42
3.3.3 组件属性动态绑定数据 42
3.3.4 控制属性动态绑定数据 43
3.3.5 关键字动态绑定数据 43
3.3.6 运算 43
3.4 微信小程序WXSS样式渲染 44
3.4.1 尺寸单位 44
3.4.2 样式导入 44
3.4.3 内联样式 45
3.4.4 选择器 45
3.4.5 常用样式属性 45

3.5 微信小程序条件渲染 49
3.5.1 使用wx: if 判断单个组件 49
3.5.2 使用block wx: if 判断多个
组件 49
3.6 微信小程序列表渲染 49
3.6.1 使用wx: for 列表渲染单个
组件 49
3.6.2 使用block wx: for 列表渲染
多个组件 50
3.6.3 使用wx: key 指定唯一
标识符 50
3.7 项目实战：任务4——实现"我
的"界面列表式导航功能（1）... 51
3.8 小结 54

第4章
莫凡商城首页静态布局
设计 55
4.1 首页需求分析与知识点 55
4.2 视图容器组件在首页中的
应用 55
4.2.1 view视图容器组件 56
4.2.2 scroll-view可滚动视图容器
组件 57
4.2.3 swiper滑块视图容器组件 59
4.2.4 movable-view可移动视图
容器组件 62
4.2.5 cover-view覆盖原生组件的
视图容器组件 64
4.2.6 项目实战：任务9——实现搜索
区域布局与海报轮播功能 65
4.3 基础内容组件 68
4.3.1 icon图标组件 68
4.3.2 text文本组件 70

4.3.3　progress 进度条组件 71
　　4.3.4　rich-text 富文本组件 71
　　4.3.5　editor 富文本编辑器及 API ... 73
4.4　image 图片组件及图片 API 的
　　　应用 76
　　4.4.1　image 图片组件 76
　　4.4.2　图片 API 的应用 79
　　4.4.3　项目实战：任务 10（1）——实现
　　　　　图书列表显示功能静态布局.... 84
4.5　导航组件和导航 API 88
　　4.5.1　navigator 页面链接组件 88
　　4.5.2　wx.navigateTo 保留当前页
　　　　　跳转 API 90
　　4.5.3　wx.redirectTo 关闭当前页
　　　　　跳转 API 91
　　4.5.4　wx.switchTab 跳转到 tabBar
　　　　　页面 API 92
　　4.5.5　wx.navigateBack 返回
　　　　　上一页 API 93
　　4.5.6　wx.reLaunch 关闭所有页面，
　　　　　打开某个页面 API 94
　　4.5.7　导航条 API 94
　　4.5.8　Tab Bar 标签导航 API 96
　　4.5.9　项目实战：任务 11——实现
　　　　　图书搜索功能 98
4.6　项目实战：任务 12——实现图书
　　　更多列表显示功能静态布局...... 102
4.7　小结 107

第 5 章
莫凡商城首页动态绑定
设计 108

5.1　微信小程序函数处理 108
　　5.1.1　生命周期函数 108
　　5.1.2　页面事件函数 110
　　5.1.3　页面路由管理 110
　　5.1.4　自定义函数 111
　　5.1.5　setData 设值函数 112
5.2　微信小程序网络请求 113
　　5.2.1　网络访问配置 113
　　5.2.2　wx.request 请求数据 API ... 116
　　5.2.3　wx.uploadFile 文件
　　　　　上传 API 118
　　5.2.4　wx.downloadFile 文件
　　　　　下载 API 120
　　5.2.5　WebSocket 会话 API 121
　　5.2.6　项目实战：任务 10（2）——实现
　　　　　图书列表显示功能动态渲染... 124
5.3　微信小程序定义模板 126
　　5.3.1　定义模板 126
　　5.3.2　使用模板 126
5.4　微信小程序的引用功能......... 127
　　5.4.1　import 引用 127
　　5.4.2　include 引用 127
5.5　WXS 小程序脚本语言........... 127
　　5.5.1　模块化 128
　　5.5.2　变量与数据类型 129
　　5.5.3　注释 130
　　5.5.4　语句 131
5.6　下拉刷新及窗口设置............. 132
　　5.6.1　下拉刷新 API 及事件 132
　　5.6.2　wx.setBackgroundColor
　　　　　动态设置窗口的背景色........ 134
　　5.6.3　wx.setBackgroundTextStyle
　　　　　动态设置下拉背景字体......... 135
　　5.6.4　wx.loadFontFace 引入第三方
　　　　　字体 136
　　5.6.5　wx.pageScrollTo 将页面
　　　　　滚动到目标位置 137
5.7　小结 137

第6章

莫凡商城的注册、登录功能 138

- 6.1 微信小程序表单组件 138
  - 6.1.1 button 按钮组件 138
  - 6.1.2 checkbox 多选项目组件 141
  - 6.1.3 radio 单选项目组件 143
  - 6.1.4 input 输入框组件 143
  - 6.1.5 textarea 多行输入框组件 146
  - 6.1.6 label 改进表单可用性组件 ... 148
  - 6.1.7 picker 滚动选择器组件 149
  - 6.1.8 slider 滑动选择器组件 158
  - 6.1.9 switch 开关选择器组件 160
  - 6.1.10 form 表单组件 162
  - 6.1.11 项目实战：任务 2——实现注册功能 163
- 6.2 微信小程序界面交互 API 167
  - 6.2.1 wx.showToast/wx.hideToast 显示/隐藏消息提示框 API 168
  - 6.2.2 wx.showModal 显示模态对话框 API 169
  - 6.2.3 wx.showLoading/wx.hideLoading 显示/隐藏 loading 提示框 API 170
  - 6.2.4 wx.showActionSheet 显示操作菜单 API 170
- 6.3 定时器 API 171
- 6.4 数据缓存 API 的获取 172
  - 6.4.1 将数据缓存到本地 173
  - 6.4.2 获取本地缓存数据 174
  - 6.4.3 清理本地缓存数据 177
  - 6.4.4 从缓存获取图书列表数据 178
- 6.5 登录相关 API 179
  - 6.5.1 登录 API 179
  - 6.5.2 获取账号信息 API 181
  - 6.5.3 获取用户信息 API 181
  - 6.5.4 授权 API 182
  - 6.5.5 设置 API 184
- 6.6 项目实战：任务 3——实现登录功能 185
- 6.7 项目实战：任务 4——实现"我的"界面列表式导航功能（2）.... 191
- 6.8 项目实战：任务 5——实现修改密码功能 195
- 6.9 项目实战：任务 6——实现意见反馈功能 199
- 6.10 项目实战：任务 7——实现清除缓存功能 201
- 6.11 小结 202

第7章

莫凡商城商品详情页设计 203

- 7.1 页面间传递数据 203
- 7.2 媒体组件及媒体 API 的应用 205
  - 7.2.1 audio 音频组件及音频 API 205
  - 7.2.2 video 视频组件及视频 API 208
  - 7.2.3 camera 相机组件及相机 API 210
  - 7.2.4 live-player 实时音视频播放 211
  - 7.2.5 live-pusher 实时音视频录制 212
- 7.3 项目实战：任务 13——实现商品详情页功能 213
- 7.4 项目实战：任务 14——实现

　　　　商品加入购物车功能 221
7.5　项目实战：任务 15——实现
　　　购物车列表功能 224
7.6　商品详情页分享与转发 API 的
　　　应用 228
7.7　小结 229

第 8 章

莫凡商城获取收货地址功能设计 230

8.1　位置 API 230
　　8.1.1　查看位置、获得位置、打开
　　　　　位置 230
　　8.1.2　开启/停止接收位置信息 232
　　8.1.3　监听实时地理位置 232
8.2　收货地址 API 233
8.3　地图组件及地图 API 234
　　8.3.1　map 地图组件 234
　　8.3.2　地图 API 的应用 237
8.4　项目实战：任务 16——实现
　　　商品立即购买页功能 238
8.5　项目实战：任务 17——实现
　　　收货地址列表功能 244
8.6　项目实战：任务 18——实现
　　　新增和编辑地址功能 247
8.7　小结 252

第 9 章

莫凡商城支付功能及订单详情页设计 253

9.1　支付 API 253

9.2　项目实战：任务 19——实现
　　　支付功能 254
9.3　画布组件及画布 API 的应用 ... 257
9.4　项目实战：任务 20——实现
　　　支付完成页功能 261
9.5　项目实战：任务 8——实现
　　　我的订单功能 262
9.6　项目实战：任务 21——实现
　　　订单详情页功能 269
9.7　小结 276

第 10 章

小程序扩展应用 277

10.1　设备应用 API 277
　　10.1.1　获得系统信息 277
　　10.1.2　获取网络状态 278
　　10.1.3　加速度计 279
　　10.1.4　罗盘 279
　　10.1.5　拨打电话 280
　　10.1.6　扫码 280
　　10.1.7　剪贴板 281
　　10.1.8　蓝牙 281
　　10.1.9　屏幕亮度 282
　　10.1.10　震动 282
　　10.1.11　手机联系人 282
10.2　文件操作 API 283
　　10.2.1　wx.saveFile 保存文件到
　　　　　本地 283
　　10.2.2　wx.getSavedFileList 获取
　　　　　本地文件列表 284
　　10.2.3　wx.getSavedFileInfo 获取
　　　　　本地文件信息 284
　　10.2.4　wx.removeSavedFile 删除
　　　　　本地文件 285

10.2.5 wx.openDocument 打开文档 285
10.2.6 wx.getFileInfo 获取文件信息 286
10.3 窗口 API 286
10.4 微信运动 API 286
10.5 项目实战：任务 22——实现图书分类功能 287
10.6 项目实战：任务 23——实现图书分类结果列表功能 290
10.7 小结 293

第1章
微信小程序概述

微信小程序自从 2017 年 1 月 9 日正式发布后，深受很多企业和开发人员的关注，很多企业将微信小程序作为公司业务的流量入口，很多试点的业务也是先做成一款小程序在市场中运行，如果业务开展顺利、有发展前景，再做 Android 版本、iOS 版本的 App。这种形式，既可以让企业尝试新业务，也可以让开发成本得到有效的控制。小程序除了在公司层面得到极大的关注外，在开发人员中间也成为热烈讨论的话题。微信小程序的开发门槛较低，很多开发人员都尝试自己来做小程序。微信小程序社区也很活跃，小程序的产品不断丰富，再加上腾讯微信小程序允许以个人身份发布，这些都给微信小程序的发展注入了新的活力。

1.1 微信小程序介绍

1.1.1 初识微信小程序

微信小程序是腾讯公司推出的在微信服务号、订阅号、企业号之后深受大众喜爱的一款新产品。这 4 种产品分别有不同的作用和功能。

（1）服务号：为企业和组织提供更强大的业务服务与用户管理能力，主要偏向服务类交互，也是我们常说的微信公众号。

（2）订阅号：为媒体和个人提供一种新的信息传播方式，主要功能是在微信侧给用户传达资讯。

（3）企业号：是企业的专业办公管理工具，具有与微信一致的沟通体验，提供丰富、免费的办公应用，并与微信消息、小程序、微信支付等互通，助力企业高效办公和管理。

（4）小程序：是一种新的开放能力，开发者可以快速地开发一个小程序。小程序可以在微信内被便捷地获取和传播，同时具有出色的使用体验。

微信小程序是一种无须安装即可使用的应用，它实现了应用"触手可及"的梦想，用户扫一扫或者搜一下即可打开应用；也体现了"用完即走"的理念，用户不再需要关心是否安装太多应用的问题。应用将无处不在，随时随地可用，无须卸载。

微信小程序也是基于去中心化而存在的一个平台，它没有聚合的入口，那么它的入口在哪里呢？在哪里可以找到小程序呢？

（1）在微信中的"发现"界面，可以找到小程序的入口，如图 1.1（a）所示。

（2）在微信主界面的下拉窗口里可以找到用过的小程序，如图 1.1（b）所示。

（3）可以直接打开好友或者群里分享的小程序，如图 1.1（c）所示。

（4）通过扫描二维码可以进入小程序，如图 1.2 所示。

微信小程序的日使用用户数超过 1.7 亿人，总数超过 58 万人。小程序不仅在一、二线城市被接受，而且在三、四线及以下城市的覆盖率达到 50%。小程序开发者已经超过 100 万人，小程序第三方平台数也已经超过 2 300 家。

（a）"发现"界面　　　　（b）主界面下拉窗口　　　　（c）分享的小程序

图 1.1　微信小程序入口

图 1.2　微信小程序二维码

1.1.2　微信小程序的功能

用户使用微信小程序能做什么呢？它能为我们提供哪些功能呢？

（1）分享页功能：用户可以将小程序当前页面分享给好友，如分享北京到上海的火车票列表界面，用户打开时是这个页面的实时数据，而不需要再次启动微信小程序。

（2）分享对话功能：可以将对话分享给好友或者分享到微信群。

（3）线下扫码进入微信小程序功能：提示用户附近有哪些微信小程序可以使用，扫描二维码就可以使用微信小程序。

（4）挂起状态功能：例如，来电话可以先接电话，接完电话后可以继续使用微信小程序进行相关操作。

（5）消息通知功能：商户可以发送消息给接受过服务的用户，用户也可以使用微信小程序的客服功能联系商户。

（6）实时音视频录制播放功能：通过此功能可以随时随地进行直播或者录播。

（7）硬件连接功能：通过使用 NFC（Near Field Communication，近距离无线通信）功能，可以把手机变成公交卡、门禁卡等进行便捷使用；通过 Wi-Fi 连接功能，进行网络连接。

（8）小游戏功能：微信小程序制作的"跳一跳"小游戏，让游戏大门从此打开，让用户知道小程序也可以制作小游戏。

（9）公众号关联功能：微信小程序可与公众号进行关联，公众号可关联不同主体的3个小程序，可关联同一主体的10个小程序，同一个小程序最多可关联3个公众号。

1.1.3 微信小程序的使用场景

从微信小程序上线开始，各种小程序就如雨后春笋般地出现。那么小程序有哪些适合的使用场景呢？在发布小程序的时候，要选择服务类目，通过这些服务类目，能知道小程序的使用场景。服务类目分为个人服务类目和企业服务类目。

- 个人服务类目的小程序开发主体为个人，它的服务类目少一些、服务范围小一些，主要包括出行与交通、生活服务、餐饮、旅游、商业服务、快递业与邮政、教育、工具、体育等。
- 企业服务类目的小程序开发主体是企业，它的服务类目多一些、服务范围大一些，主要包括快递业与邮政、教育、医疗、政务民生、金融业、出行与交通、房地产、生活服务、IT科技、餐饮、旅游、时政信息、文娱、工具、电商平台、商家自营、商业服务、公益、社交、体育、汽车等。

1.1.4 微信小程序的发展历程

微信小程序从开始研发到正式发布，经历了一年的时间。

（1）2016年1月9日，微信团队首次提出应用号的概念。

（2）2016年9月22日，"微信公众平台"对外发送小程序内测邀请，内测名额200个。

（3）2016年11月3日，微信小程序对外公测，开发完成后可以提交审核，但公测期间不能发布。

（4）2016年12月28日，微信创始人张小龙在微信公开课中解答外界对微信小程序的几大疑惑，包括没有应用商店、没有推送消息等。

（5）2016年12月30日，"微信公众平台"对外公告，上线的微信小程序最多可生成10 000个带参数的二维码。

（6）2017年1月9日，微信小程序正式上线。

（7）2017年2月，开发模糊搜索、摩拜单车小程序，接入微信扫一扫。

（8）2017年3月27日，个人开发者可以申请小程序的开发和发布，通过公众号菜单、模板消息可打开小程序，关联小程序可下发通知，App可分享小程序，小程序兼容线下已有二维码。

（9）2017年4月17日，小程序代码包的大小限制扩大到2 MB，提供小程序码，开放小程序第三方平台，公众号文章可进入小程序，提供门店小程序。

（10）2017年4月20日，发布公众号关注小程序新规则。

（11）2017年5月，发布"小程序数据助手"，页面新增"转发"按钮，提供附件的小程序功能。

（12）2017年9月，微信搜索框新增小程序入口，支付后可关联公众号。

（13）2017年12月，微信更新的6.6.1版本开放了小游戏，代码包扩大到4 MB，升级实时录播视频及播放能力，升级小程序任务栏菜单。

（14）2018年1月18日，微信提供了电子化的侵权投诉渠道，用户或者企业可以在"微信公众平台"及微信客户端入口进行投诉。

（15）2018年1月25日，微信团队在"微信公众平台"发布公告称，从移动应用分享至微信的小程序页面，用户访问时支持打开来源应用。

（16）2018年3月，微信正式宣布小程序广告组件启动内测，内容还包括第三方可以快速创建并认证小程序、新增小程序插件管理接口和更新基础能力，开发者可以通过小程序来赚取广告收入，开放对个人开发者的使用权限。

（17）2018年4月，通过公众号文章可以打开小程序，开放微信小程序游戏接口。

（18）2018年5月，支持App打开小程序。

（19）2018年6月，小程序支持打开公众号文章（关联的公众号），更新开发者工具：代码云托管，优化预览方式和界面布局，代码包扩大到8 MB。

（20）2018年7月，开放品牌搜索功能，推出品牌官方区和微主页，任务栏出现"我的小程序"入口（iOS：7月13日）。

（21）2018年8月，微信小程序云开发上线，支持iPad打开小程序。

（22）2018年10月，小程序支持主体迁移。

（23）2019年8月，微信向开发者发布新能力公测与更新公告，微信PC版新版本支持打开聊天中分享的小程序。

1.1.5 微信小程序带来的机会

截至2018年年底，微信小程序覆盖了超过200个细分行业，服务了超过1 000亿人次用户；城市服务覆盖了362个地级城市，年交易增长量超过了600%，创造了超过5 000亿元的商业价值。在微信小程序整体用户中，男性占58%，40岁以下用户占70%；76%的用户学历在大专以上，三、四线及以下城市用户占比分别为22%、30%。71%的用户分享过小程序，其中14%的用户主动分享意愿高；分享内容的价值性成为用户判断分享与否的首要标准，57%的用户会分享有价值的内容；此外，朋友要求、利益诱导也是分享的考虑因素。

微信团队在小程序开发方面也给了小程序开发者很多支持，这给开发者带来了很大的机会，因为微信小程序的门槛很低，不需要太难的技术，学习微信小程序，就可以成为一名"小程序员"，如设计师、学生、创业者、待业青年、"网虫"、策划人员、编辑、"草根"站长等都可以转做程序员。

微信团队大力扶持开发者，未来将提供与服务商相关的更多能力，为普通商户和服务商牵线搭桥。微信小程序给企业提供了流量入口，企业可以通过小程序推广自己的产品，从iOS、Android、网站、公众号入口到新增小程序入口。由于腾讯公司的大力扶持，小程序也成为各个企业非常看重的流量入口。

1.2 微信小程序环境搭建

慕课视频

微信小程序环境搭建

1.2.1 小程序环境搭建

微信小程序环境搭建很简单，需要3个步骤。

（1）下载微信开发者工具。在"微信公众平台"官网选择"小程序"→"小程序开发文档"，如图1.3所示。在打开的界面中选择"工具"→"下载"，可以看到微信小程序为不同的操作系统（Windows 64、Windows 32、macOS）提供了不同版本的开发者工具（因为开发者工具的版本更新很快，实际版本以读者操作为准），我们可以根据自己的操作系统下载相应的版本，如图1.4所示。

（2）注册"微信公众平台"账号。在"微信公众平台"官网选择微信小程序类别进行注册。小程序支持个人开发者账号，可以注册个人账号；如果公司需要小程序，也可以进行企业认证。通过这两种方式都可以发布小程序，如图1.5所示。

图 1.3 小程序开发文档

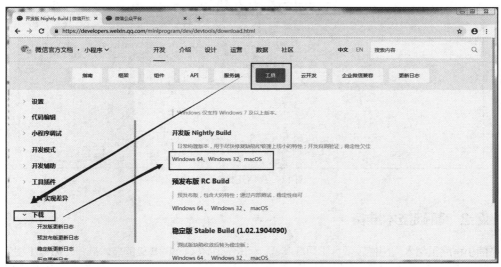

图 1.4 下载微信开发者工具

图 1.5 注册"微信公众平台"账号

（3）使用下载的微信开发者工具包。按照提示完成开发者工具的安装，安装完成后，运行微信开发者工具，会出现一个二维码，需要用绑定"微信公众平台"的微信扫描登录，登录后可以发现，开发者工具提供了本地"小程序项目"开发和"公众号网页项目"开发两个调试类型。在小程序开发部分微信开发者工具支持新建小程序和导入已有的小程序，如图1.6所示。

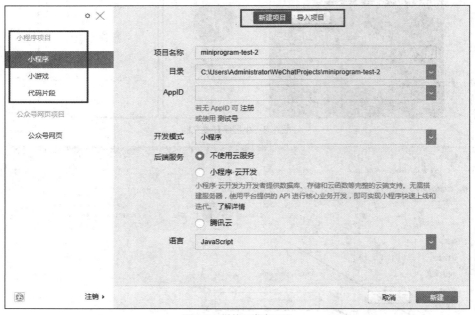

图1.6 微信开发者工具

1.2.2 基础技术准备

微信小程序虽然入门门槛较低、学习成本低，但是它也需要一些基础的技术准备。微信小程序自定义了一套语言，称为 WXML 微信标记语言，它的使用方法类似于 HTML 语言，所以我们需要对 HTML 语言有所了解；微信小程序还定义了自己的样式语言 WXSS，它兼容了 CSS（层叠样式表）样式，并做了扩展，所以我们需要对 CSS（层叠样式表）有所了解；微信小程序是使用 JavaScript 来进行业务处理的，兼容了大部分 JavaScript 功能，但仍有一些功能无法使用，所以我们还需要对 JavaScript 语言有所了解。有 HTML、CSS 和 JavaScript 技术功底的人学习微信小程序的开发，就非常容易了。

1.3 微信小程序开发者工具的使用

在进行小程序开发之前，需要先掌握微信开发者工具的使用，对工具的使用掌握得越熟练，越有助于小程序的开发和提高开发效率。下面介绍一下微信开发者工具的使用。

1.3.1 如何创建项目

创建一个小程序项目，需要准备 AppID、项目名称和项目目录。

（1）获取微信小程序 AppID，需要在"微信公众平台"中登录1.2.1小节中注册的账号，在"开发"→"开发设置"中，查看微信小程序的 AppID，如图1.7所示。

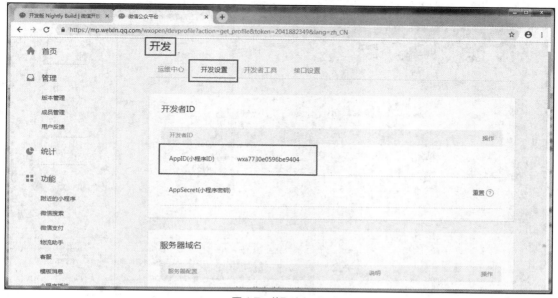

图 1.7 获取 AppID

（2）创建一个"小程序开发工具使用"项目，项目文件存放到"chapter1"文件夹中，其他都使用默认配置，如图 1.8 所示。

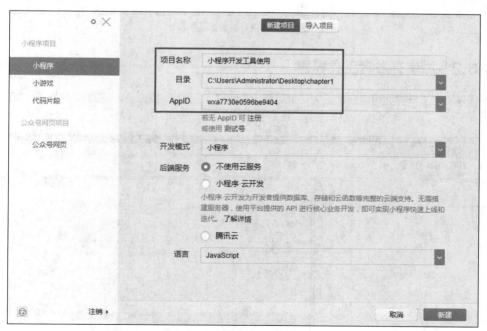

图 1.8 创建小程序项目

（3）单击"新建"按钮，即可创建一个小程序项目，同时进入微信开发者工具界面，如图 1.9 所示。

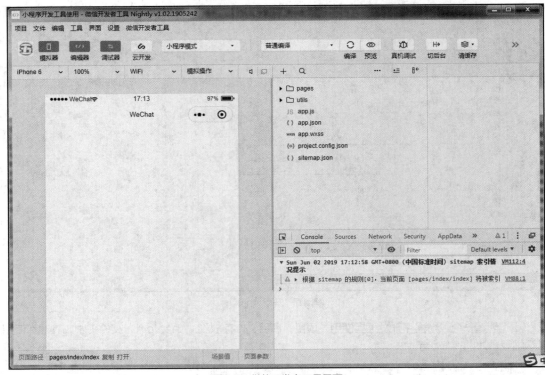

图 1.9 微信开发者工具界面

1.3.2 微信开发者工具界面

微信开发者工具界面可以分为五大功能区域：菜单栏、工具栏、模拟器、编辑器、调试器，如图 1.10 所示。

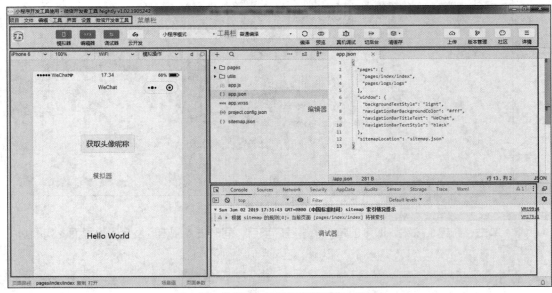

图 1.10 微信开发者工具界面的五大功能区域

1. 菜单栏

菜单栏有项目、文件、编辑、工具界面、设置、微信开发者工具 7 个菜单项，涉及软件的一些常规操作和功能使用。

"项目"菜单：通过"项目"菜单，可以新建项目、导入项目、打开最近项目、新建代码片段、导入代码片段、查看所有项目和关闭当前项目，从而对小程序项目或者代码片段进行管理和使用。

"文件"菜单：通过"文件"菜单，可以新建文件、保存文件、保存所有文件和关闭文件。

"编辑"菜单：是对代码进行管理的一个菜单，通过该菜单可以格式化代码等。

"工具"菜单：用于编译、刷新、预览、清除缓存等。

"界面"菜单：用于显示或者隐藏工具栏、模拟器、编辑器、目录树、调试器区域。

"设置"菜单：通过"设置"菜单，可以进行通用设置、外观设置、快捷键设置、编辑设置、代理设置、安全设置、项目设置。

"微信开发者工具"菜单：通过该菜单，可以对开发者工具进行升级、回退和退出等。

2. 工具栏

工具栏包含了小程序开发中使用到的常用工具和常用功能。

（1）显示或隐藏模拟器、编辑器、调试器按钮。这 3 个按钮可以控制模拟器、编辑器、调试器区域的显示或者隐藏。

（2）"云开发"按钮。通过"云开发"按钮可以进入小程序云开发控制台，进行小程序云开发。

（3）小程序模式。这里提供两种模式：一种是小程序模式，用于正常开发小程序项目；另一种是插件模式，用于开发小程序插件。

（4）编译操作。可以通过使用"编译"按钮或者"Ctrl + B"组合键，编译当前代码，并自动刷新模拟器，同时鉴于开发者需调试从不同场景值进入具体的页面，开发者可以添加或选择已有的自定义编译条件进行编译和代码预览，如图 1.11 所示。

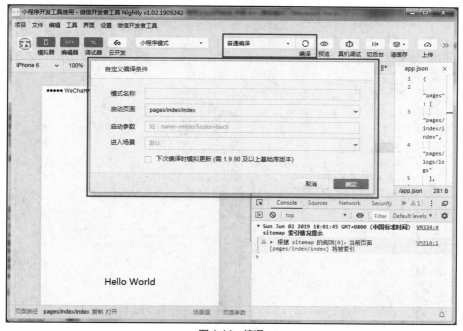

图 1.11　编译

（5）预览。单击"预览"按钮，可以将小程序上传，生成二维码，通过扫描二维码可以在手机上预览小程序，如图1.12所示。

图1.12 预览（图中二维码仅为示意，请扫描自己操作生成的二维码）

（6）真机调试。小程序允许在真机上调试，可以在发布之前查看小程序在真机上的运行效果。

（7）前、后台切换。工具栏中前、后台切换按钮可帮助开发者模拟一些客户端的环境操作。例如，在操作微信小程序的过程中，突然有电话进来，如果接电话，小程序就会从前台进入后台，重新访问小程序时，小程序又会从后台进入前台，如图1.13所示。

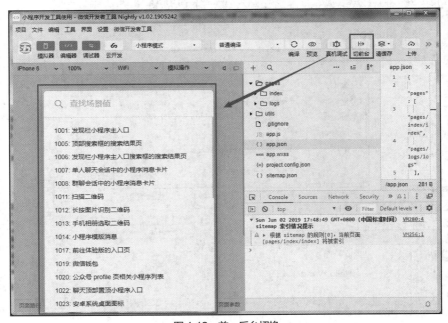

图1.13 前、后台切换

（8）清缓存。清缓存包括清除数据缓存、清除文件缓存、清除授权数据、清除网络缓存、清除登录状态、全部清除功能，如图 1.14 所示。

（9）上传。小程序开发完成后需要上传到腾讯服务器进行测试如图 1.15 所示。在手机上可以预览开发版本的小程序。

（10）版本管理。小程序代码可以上传到 Git 进行版本管理和多人协作开发。

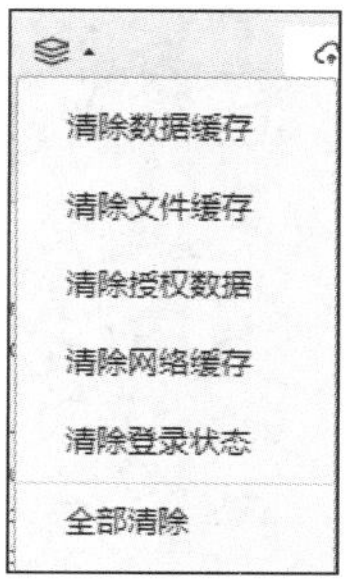

3. 模拟器

模拟器区域用来显示小程序界面。在小程序开发过程中，小程序界面会随着代码的编写实时变化，以方便小程序的开发和调试；同时，模拟器可以模拟小程序在各个终端设备上的操作效果，可以设置小程序运行的终端设备，如"iPhone 5""iPhone 6"等终端设备，可以设置模拟器区域的百分比大小，可以模拟设置网络连接"Wi-Fi""2G""3G"等网络情况，如图 1.16 所示。

图 1.14　清缓存

图 1.15　上传

图 1.16　模拟器

4. 编辑器

编辑器区域分为两部分：一部分是用来展示项目文件目录和文件结构的，称为目录树；另一部分是用来编辑代码的区域，如图 1.17 所示。

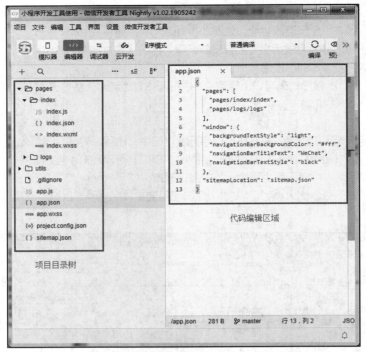

图 1.17 编辑器区域

（1）在项目目录上单击鼠标右键，在弹出的快捷菜单中可以新建目录、Page、Component、JS、TS、JSON、WXML、WXSS、WXS，可以对文件目录重新命名，可以删除目录，可以在该目录下查找指定内容，如图 1.18 所示。

图 1.18 文件操作

（2）在代码编辑区域中编写代码后，可以通过模拟器区域，实时预览编辑的小程序的演示情况。修改 WXSS、WXML 文件，会刷新当 Page，修改 JS、JSON 文件，会重新编译小程序，如图 1.19 所示。

（3）在代码编写过程中，开发者工具提供自动补全功能。在编辑 JS 文件时，开发者工具会帮助开发者补全所有的 API（Application Programming Interface，应用程序编程接口），并给出相关的注释解释；在编辑 WXML 文件时，开发者工具会帮助开发者直接写出相关的标签；在编辑 JSON 文件时，会帮助开发者补全相关的配置，并给出实时的提示，如图 1.20 所示。

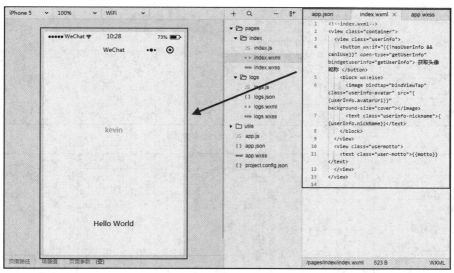

图 1.19　代码编写

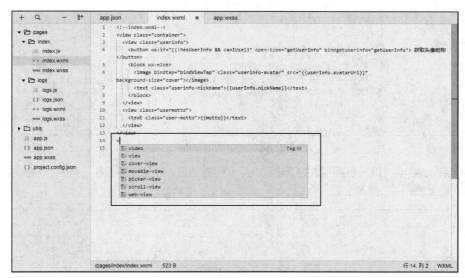

图 1.20　自动补全功能

（4）开发者工具提供自动保存功能。书写代码后，工具会自动帮助用户保存当前的代码编辑状态，直接关闭工具或者切换到别的项目，并不会丢失已经编辑的文件内容，但需要注意的是，只有保存文件，修改内容才会真实地写到硬盘上，并触发实时预览。

5．调试器

小程序常用的调试工具有 Console、Sources、Network、Storage、AppData、Wxml。

（1）Console 窗口用来显示小程序的错误信息和调试代码，还可以进行代码编写和调试，如图 1.21 所示。

（2）Sources 窗口用于显示当前项目的脚本文件，在 Sources 中开发者看到的文件是经过处理之后的脚本文件，开发者的代码都会被包裹在 define 函数中，并且对于 Page 代码，可通过 require 主动调用，如图 1.22 所示。

图 1.21　Console 的功能

图 1.22　Sources 的功能

（3）Network 窗口用来观察发送的请求和调用文件的信息，包括文件名称、路径、大小、调用的状态、时间等，如图 1.23 所示。

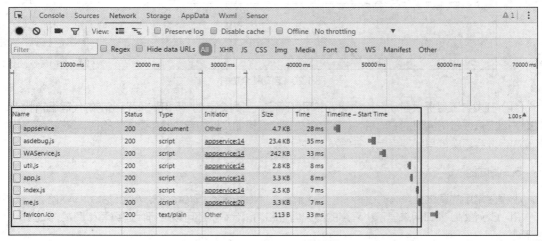

图 1.23　Network 的功能

（4）Storage 窗口用于显示当前项目使用 wx.setStorage 或者 wx.setStorageSync 后的数据存储情况，如图 1.24 所示。

图1.24　Storage 的功能

（5）AppData 窗口用于显示当前项目当前时刻的具体数据，实时地反馈项目数据情况，开发者可以在此处编辑数据，并将其及时地反馈到界面上，如图1.25所示。

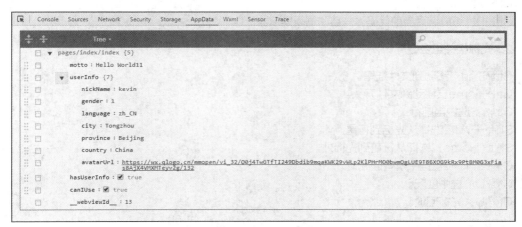

图1.25　AppData 的功能

（6）Wxml 窗口用于帮助开发者开发 Wxml 转化后的界面。开发者在这里可以看到真实的页面结构及结构对应的 WXSS 属性，同时可以修改对应的 WXSS 属性，如图1.26所示。

图1.26　Wxml 的功能

1.3.3 常用快捷键

1. 格式调整组合键

Ctrl+S：保存文件。
Ctrl+[，Ctrl+]：代码行缩进。
Ctrl+Shift+[，Ctrl+Shift+]：折叠打开的代码块。
Ctrl+C Ctrl+V：复制粘贴，如果没有选中任何文字则复制粘贴一行。
Shift+Alt+F：代码格式化。
Alt+Up，Alt+Down：上、下移动一行。
Shift+Alt+Up，Shift+Alt+Down：向上、向下复制一行。
Ctrl+Shift+Enter：在当前行上方插入一行。
Ctrl+Shift+F：全局搜索。
Ctrl+B：可以编译当前代码，并自动刷新模拟器。

2. 光标相关组合键

Ctrl+End：移动到文件结尾。
Ctrl+Home：移动到文件开头。
Ctrl+i：选中当前行。
Shift+End：选择从光标到行尾。
Shift+Home：选择从行首到光标处。
Ctrl+Shift+L：选中所有匹配。
Ctrl+D：选中匹配。
Ctrl+U：光标回退。

3. 界面相关快捷键

Ctrl + \：隐藏侧边栏。
Ctrl + m：打开或者隐藏模拟器。

1.4 项目实战：创建莫凡商城小程序

项目实战：创建莫凡商城小程序

莫凡商城（mofunShop）小程序是贯穿本书的项目案例，通过学习本书内容可以完整地开发一个企业级的小程序。本章介绍如何创建微信小程序项目及如何使用微信开发者工具。下面开始创建一个莫凡商城小程序。

（1）打开微信开发者工具，创建莫凡商城小程序，并将其存放到"mofunShop"文件夹，AppID 使用自己在公众平台里的 AppID，如图 1.27 所示。

（2）创建莫凡商城小程序项目之后，会进入到微信开发者工具里，默认创建一个小程序页面，包括两方面内容：一是输出 Hello World 文字；二是获取用户的头像、昵称信息。在 pages/index/index.js 文件里，Page 的 data 里提供数据源 motto，data 的数据可以动态地绑定到 WXML 页面里，如图 1.28 所示。

（3）在 pages/index/index.wxml 文件里，通过双大括号（{{}}）的方式，将 motto 绑定到页面里，motto 对应的值就可以在页面中显示出来，如图 1.29 所示。

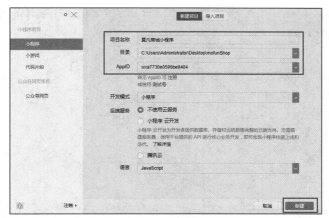

图 1.27 创建项目

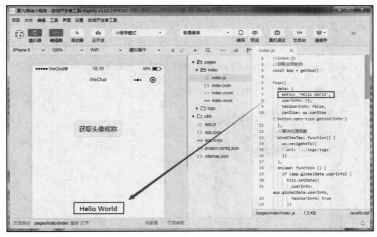

图 1.28 显示 Hello World

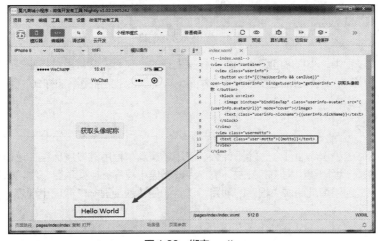

图 1.29 绑定 motto

（4）在 pages/index/index.wxss 文件里，通过 class 的方式给 Hello World 添加样式，使其距顶部的高度为 200 px，如图 1.30 所示。

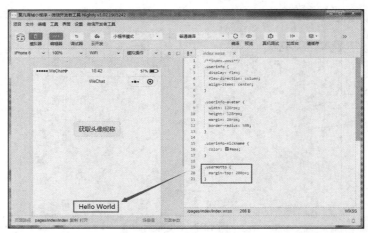

图 1.30　添加样式

（5）单击"获取头像昵称"按钮，可以通过触发获取用户信息事件（getUserInfo）来获取用户的相关信息。微信小程序获取用户头像、昵称，需要用户授权后方可渲染到界面上，如图 1.31 所示。

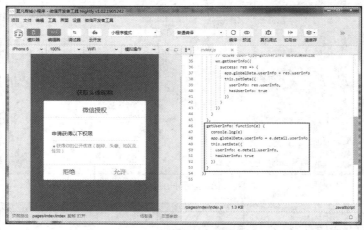

图 1.31　获取头像昵称

在实际的开发过程中，小程序的编写也是这样来进行的：在 JS 文件里进行业务逻辑的处理，动态地提供数据；在 WXML 文件里绑定数据，渲染界面；在 WXSS 文件里添加样式，美化页面。

1.5　小结

本章主要介绍微信小程序是什么、能干什么、使用场景及小程序的发展历程，让读者对小程序有一个初步的认识；带领读者搭建小程序的开发环境，完成微信小程序的创建项目，学习微信小程序相关的基础知识；最后完成创建莫凡商城小程序的项目，体验如何快速创建一个微信小程序。

第2章
莫凡商城小程序项目任务

莫凡商城小程序是一个完整的小程序项目,主要实现图书商品的展示与销售功能,是非常具有代表性的小程序应用项目。小程序分为 4 个功能模块:"我的"模块、"首页"模块、"购买商品"模块和图书"分类"模块。"我的"模块用来显示用户的订单,并实现修改密码、意见反馈等功能,如图 2.1 所示。"首页"模块显示海报轮播图、"热门技术"图书区域、"秒杀时刻"图书区域、"畅销书籍"图书区域,如图 2.2 所示;"购买商品"模块中的"购物车"就像我们常用的一样,用于显示想要购买的商品列表,如图 2.3 所示;"图书分类"模块将要销售的图书进行一、二级分类,并分类显示图书商品,如图 2.4 所示。

　　图 2.1　我的　　　　　图 2.2　首页　　　　　图 2.3　购物车　　　　图 2.4　图书分类

本章针对莫凡商城小程序进行功能拆解,将项目拆解成一个个任务,以便在后续章节里学习小程序的过程中,逐步实现这些功能,各个击破,最后形成一个完整的小程序项目。莫凡商城小程序的项目任务具有代表性,在设计其他小程序的时候,可以参照莫凡商城小程序的创建。

2.1　"我的"模块功能介绍

"我的"模块实现底部标签导航功能、注册功能、登录功能、"我的"界面列表式导航功能、修改密码功能、意见反馈功能、清除缓存功能、我的订单功能。

2.1.1　任务 1——实现底部标签导航功能

莫凡商城小程序的底部标签导航包括"首页""分类""购物车"和"我的"4 个标签。切换底部标签,可以显示对应的导航内容,如图 2.5 所示。

设计目的：底部标签导航是绝大多数小程序都会应用到的一种设计方式，其功能的实现，对设计其他小程序具有借鉴意义。

图 2.5　底部标签导航

2.1.2　任务 2——实现注册功能

注册是 App 和小程序中不可或缺的功能，要让用户在莫凡商城中创建一个属于自己的账号，就需要用到注册功能，如图 2.6 所示。

图 2.6　注册功能

设计目的：注册功能是小程序经常会用到的功能，可以通过用户的注册，来管理用户的账号信息。通过注册表单的设计，我们可以学到表单校验、设计注册表单和提交注册表单的方法。

2.1.3　任务 3——实现登录功能

莫凡商城小程序提供了两种登录方式：账号密码登录和手机快捷登录，如图 2.7 和图 2.8 所示。

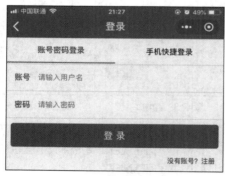

图 2.7　账号密码登录

图 2.8　手机快捷登录

设计目的：账号密码登录和手机快捷登录是现在比较流行的登录设计方式，很多 App 或者小程序都会采用这样的设计方式，其功能的实现对设计其他小程序具有借鉴意义。

2.1.4　任务 4——实现"我的"界面列表式导航功能

莫凡商城小程序中的"我的"界面采用列表式导航来显示内容，这也是很多 App 会采用的一种方式，如图 2.9 所示。

设计目的：列表式导航是比较普遍使用的一种设计方式，在"我的"页面里可能有很多菜单选项，列表式导航可以帮我们很好地进行布局。

图 2.9 "我的"界面列表式导航功能

2.1.5 任务 5——实现修改密码功能

莫凡商城小程序用户可以修改自己的密码,通过输入原密码、输入新密码、确认密码 3 步就可以完成密码的修改,如图 2.10 所示。

图 2.10 修改密码功能

设计目的:通过修改密码功能,我们要学会布局密码修改界面和表单校验的方法。

2.1.6 任务 6——实现意见反馈功能

莫凡商城小程序提供意见反馈功能,用户可以填写自己的意见或建议,如图 2.11 所示。

图 2.11 意见反馈

设计目的:意见反馈功能用来收集用户对产品使用的意见或建议,我们可以通过收集这些意见或建议来优化 App 或者小程序。

2.1.7 任务 7——实现清除缓存功能

莫凡商城小程序提供清除缓存功能，可以清理掉缓存在小程序中的数据。

设计目的：小程序可以提供清除缓存功能，把缓存到本地的数据清除掉，释放空间。

2.1.8 任务 8——实现我的订单功能

莫凡商城小程序我的订单功能提供"待付款"订单列表、"待收货"订单列表、"已完成"订单列表，如果没有订单，则显示空列表，如图 2.12～图 2.15 所示。

图 2.12　待付款列表　　图 2.13　待收货列表　　图 2.14　已完成列表　　图 2.15　空列表

设计目的：商城类 App 或者小程序都有订单这个概念，订单可以分为待付款订单、待收货订单、已完成订单等不同类别，每种订单具有不同的操作按钮，这是设计订单列表时要考虑的事情。

2.2 "首页"模块功能介绍

莫凡商城小程序首页提供海报轮播功能、图书列表显示功能、图书搜索功能、图书更多列表显示功能。

2.2.1 任务 9——实现搜索区域布局与海报轮播功能

首页上方是搜索区域和海报轮播区域。海报轮播功能是 App 和小程序很常用的功能，它利用界面有限的空间，通过轮播的形式显示不同的轮播内容，如图 2.16 和图 2.17 所示。

图 2.16　海报轮播 1　　　　　　　　　图 2.17　海报轮播 2

设计目的：首页搜索和海报轮播是比较通用的功能，海报轮播功能常用于重点产品或特殊活动的海报展示。

2.2.2 任务 10——实现图书列表显示功能静态布局与动态渲染

莫凡商城小程序在首页显示图书列表，包括"热门技术"列表、"秒杀时刻"列表、"畅销书籍"列表，如图 2.18 所示。本任务分静态布局与动态渲染两步实现。

图 2.18　图书列表显示

设计目的：首页采用列表的方式来设计是很常见的形式，在首页有限的空间里放置指定数量的商品，并且放置可以查看更多商品的入口，单击进入就可以查看更多的商品列表。

2.2.3 任务 11——实现图书搜索功能

莫凡商城小程序提供搜索功能，在搜索功能里提供搜索框和热门搜索标签列表，如图 2.19 所示，搜索结果界面如图 2.20 所示。

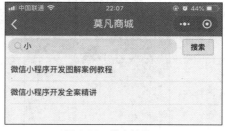

图 2.19　搜索界面及热门搜索　　　　图 2.20　搜索结果

设计目的：搜索是小程序必不可少的功能，商城里有很多商品，搜索常常是用户最先使用到的功能。

2.2.4 任务 12——实现图书更多列表显示功能

莫凡商城小程序在首页每个类别只显示 3 本书，单击"查看更多"将显示该类别的所有图书列表，如图 2.21～图 2.23 所示。

图 2.21　热门技术列表

图 2.22　秒杀时刻列表

图 2.23　畅销书籍列表

设计目的：通过制作图书更多列表可以学会商品列表的展示及不同商品分类的页签切换显示。

2.3 "购买商品"模块功能介绍

莫凡商城小程序购买商品功能包括商品详情页功能、商品加入购物车功能、商品立即购买页功能、新增地址、编辑地址、地址列表。

2.3.1 任务13——实现商品详情页功能

商品详情页用于商品轮播图片、商品介绍、商品价格、商品数量、图书详情、出版信息等内容的显示，如图 2.24 和图 2.25 所示。

图 2.24　商品详情页 1

图 2.25　商品详情页 2

设计目的：商品详情页的内容比较多、页面比较长，既要展示商品基本信息，又要有加入购物车、立即购买的操作按钮，在设计时最好将按钮固定在底部区域，不随页面的滚动而改变位置，方便用户随时将商品加入购物车或者立即购买。

2.3.2 任务14——实现商品加入购物车功能

可以将商品加入购物车里，购物车中商品的数量会变化，如图2.26所示。

图2.26 加入购物车

设计目的：加入购物车功能是比较常用的功能，本任务讲解如何设计加入购物车功能。

2.3.3 任务15——实现购物车列表功能

购物车以列表的形式显示已加入购物车的商品，可以显示选中商品的总价并进行结算，如图2.27所示。

图2.27 购物车列表

设计目的：很多商城都会设计购物车功能，本任务学习购物车功能的设计和实现。

2.3.4 任务16——实现商品立即购买页功能

在商品立即购买页面中可以看到要购买的商品名称、数量及价格，同时可以选择收货地址，如图2.28所示。
设计目的：商品立即购买页面是购买前最终确定订单的界面，在此页面可以提交订单来发起支付。

2.3.5 任务17——实现收货地址列表功能

收货地址列表用来显示用户购买商品时可以选择的收货地址，这些收货地址通过列表的方式进行显示，如图2.29所示。

设计目的：收货地址列表的展示，方便用户在下单的时候直接选择以前录入过的地址信息，这样就不用每次输入地址，方便用户快速下单。

图 2.28　商品立即购买页面　　　　图 2.29　收货地址列表

2.3.6　任务 18——实现新增和编辑地址功能

通过新增地址功能可以创建新的收货地址，对于已有的收货地址，也可以重新进行编辑，如图 2.30 和图 2.31 所示。

图 2.30　新增地址　　　　图 2.31　编辑地址

设计目的：新增地址和编辑地址是常用的功能，用以动态地维护地址信息。

2.3.7　任务 19——实现支付功能

微信小程序只支持微信支付，提供微信支付 API，莫凡商城在商品提交订单页和订单详情页都可以发起微信支付，如图 2.32 和图 2.33 所示。

图 2.32　提交订单

图 2.33　订单详情付款

设计目的：支付功能是购买商品必不可少的功能，本任务讲解如何设计微信小程序支付。

2.3.8　任务 20——实现支付完成页功能

对于商品的购买，微信小程序只能进行小程序支付，接入小程序支付，支付成功后，会跳转到支付成功页面，如图 2.34 所示。

图 2.34　支付成功页面

设计目的：支付完成后，小程序会给用户支付成功的友好提示，既提示用户支付成功，又给用户提供查看详情的入口。

2.3.9　任务 21——实现订单详情页功能

支付成功后，可以在订单详情页中查看订单相关信息、收货地址信息及商品信息，如图 2.35 所示。

设计目的：用户在购买商品的时候，有时不会立即支付，或者支付后想查看自己的订单信息，订单详情页面就是必不可少的页面了。

图 2.35　订单详情页面

2.4　"图书分类"模块功能介绍

莫凡商城小程序针对图书进行分类管理，以分类的形式显示图书相关内容。

2.4.1　任务 22——实现图书分类功能

图书分类分为一级分类和二级分类，通过手风琴式导航来显示，如图 2.36 和图 2.37 所示。

图 2.36　图书分类 1

图 2.37　图书分类 2

设计目的：手风琴式导航方式是很多商城都会采用的导航方式，对于商品分类过多的商城，这样的设计方便用户查找商品。

2.4.2　任务23——实现图书分类结果列表功能

通过图书一级分类和二级分类，可以查看到该分类下的所有图书，如图2.38所示。

图2.38　图书分类结果列表

设计目的：根据商品的分类，可以查看该分类下的商品，这样查找商品具有目的性，方便用户快速查找到想要的商品。

2.5　小结

莫凡商城小程序包含四大功能模块："我的"模块、"首页"模块、"购买商品"模块、"图书分类"模块、我们将这四大模块的功能拆分成23个任务来实现，完成莫凡商城小程序的开发，同时学会微信小程序的知识点和开发技能。

第3章 莫凡商城小程序的项目结构

在 1.4 节中，我们创建了莫凡商城小程序项目，默认生成了项目结构和文件，这些文件分为框架页面文件、工具类文件、框架全局文件 3 类，如图 3.1 所示。本章我们介绍微信小程序的项目结构。

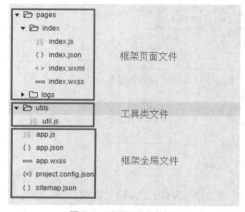

图 3.1 项目目录树结构

3.1 项目目录树结构介绍

3.1.1 框架全局文件

一个小程序的框架全局文件有 5 个：app.js 文件、app.json 文件、app.wxss 文件、project.config.json 文件、sitemap.json 文件，这 5 个文件必须放在项目的根目录中。app.js 文件是小程序的逻辑文件（定义全局数据、定义函数的文件）；app.json 文件是小程序的公共配置文件；app.wxss 文件是小程序的公共样式表文件；project.config.json 文件是个性化项目配置文件；sitemap.json 文件用于配置小程序及其页面是否允许被微信索引，它们对所有页面都有效，如表 3.1 所示。

表 3.1 框架全局文件

文件	是否必填	作用
app.js	是	编写小程序逻辑
app.json	是	进行小程序公共配置
app.wxss	否	提供小程序公共样式表
app.config.json	是	进行小程序项目个性化配置
sitemap.json	是	配置小程序及其页面是否允许被微信索引

1. app.js 小程序逻辑文件

app.js 文件用来定义全局数据和函数，它可以指定微信小程序的生命周期函数。生命周期函数可以理解为微信小程序自己定义的函数，如 onLaunch（监听小程序初始化）、onShow（监听小程序显示）、onHide（监听小程序隐藏）等，在不同阶段、不同场景可以使用不同的生命周期函数。此外，app.js 文件中还可以定义一些全局的函数和数据，其他页面引用 app.js 文件后就可以直接使用全局函数和数据，如图 3.2 所示。

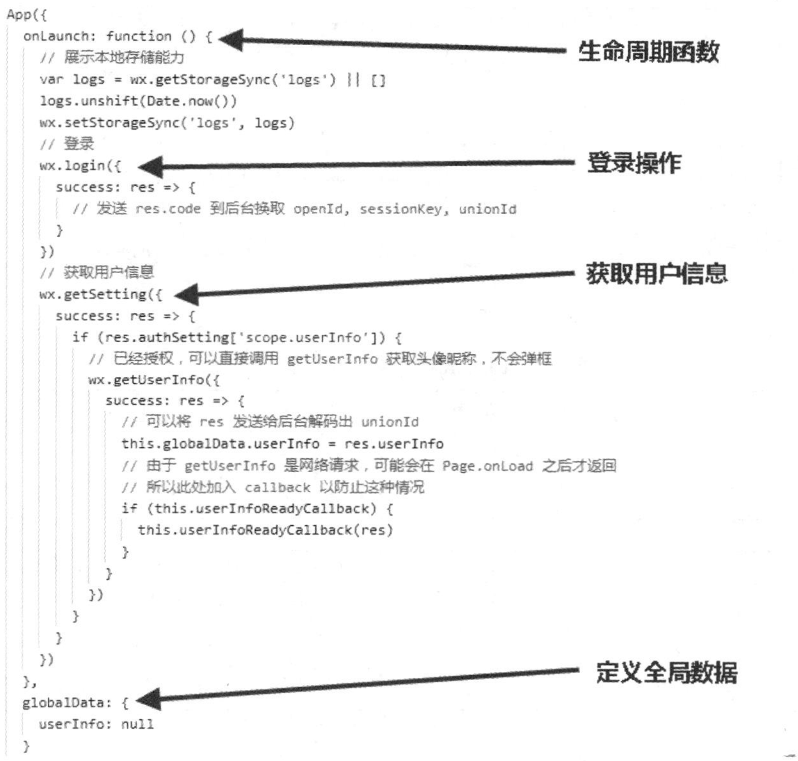

```
App({
    onLaunch: function () {           ← 生命周期函数
        // 展示本地存储能力
        var logs = wx.getStorageSync('logs') || []
        logs.unshift(Date.now())
        wx.setStorageSync('logs', logs)
        // 登录
        wx.login({                    ← 登录操作
            success: res => {
                // 发送 res.code 到后台换取 openId, sessionKey, unionId
            }
        })
        // 获取用户信息
        wx.getSetting({               ← 获取用户信息
            success: res => {
                if (res.authSetting['scope.userInfo']) {
                    // 已经授权，可以直接调用 getUserInfo 获取头像昵称，不会弹框
                    wx.getUserInfo({
                        success: res => {
                            // 可以将 res 发送给后台解码出 unionId
                            this.globalData.userInfo = res.userInfo
                            // 由于 getUserInfo 是网络请求，可能会在 Page.onLoad 之后才返回
                            // 所以此处加入 callback 以防止这种情况
                            if (this.userInfoReadyCallback) {
                                this.userInfoReadyCallback(res)
                            }
                        }
                    })
                }
            }
        })
    },
    globalData: {                     ← 定义全局数据
        userInfo: null
    }
})
```

图 3.2　app.js 小程序逻辑文件

在莫凡商城小程序里，配置接口访问域名、微信登录凭证 code 值、用户 ID，需要在 globalData 对象里配置，代码如下。

```
globalData: {
    userInfo: null,
    host: 'https://api.mofun365.com:8888',
    code: null, //微信登录凭证 code
    userId: null//用户 ID
}
```

2. app.json 小程序公共配置文件

app.json 文件用来对微信小程序进行全局配置，文件内容为一个 JSON 对象，其主要功能为配置页面路径、配置窗口表现、配置标签导航、配置网络超时、配置 debug 模式，如图 3.3 所示；另外，还可以配置是否启用插件功能页，配置分包结构，设置 Worker 代码放置的目录、需要在后台使用的能力、使用到的插件、分包预下载规则、iPad 小程序是否支持屏幕旋转、需要跳转的小程序列表、全局自定义组件、小程序接口权限、sitemap.json 文件的位置等，如表 3.2 所示。

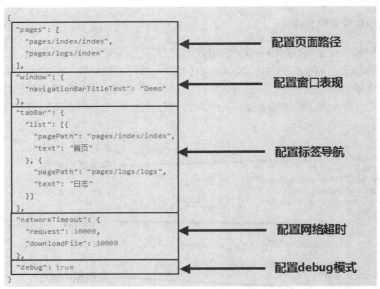

图 3.3 app.json 文件的 5 个功能

表 3.2 app.json 框架全局文件

文件	类型	是否必填	作用
pages	string[]	是	配置页面路径
window	Object	否	配置全局的默认窗口表现
tabBar	Object	否	配置底部 tab 栏的表现
networkTimeout	Object	否	配置网络超时
debug	boolean	否	设置是否开启 debug 模式,默认关闭
functionalPages	boolean	否	设置是否启用插件功能页,默认关闭
subpackages	Object[]	否	进行分包结构配置
workers	string	否	设置 Worker 代码放置的目录
requiredBackgroundModes	string[]	否	声明需要在后台使用的能力,如"音乐播放"
plugins	Object	否	声明使用到的插件
preloadRule	Object	否	声明分包预下载规则
resizable	boolean	否	设置 iPad 小程序是否支持屏幕旋转,默认关闭
navigateToMiniProgramAppIdList	string[]	否	声明需要跳转的小程序列表
usingComponents	Objec	否	进行全局自定义组件配置
permission	Object	否	进行小程序接口权限的相关设置
sitemapLocation	string	是	指明 sitemap.json 文件的位置

(1) pages: 配置页面路径。它定义一个数组,存放多个页面的访问路径,是进行页面访问的必要条件。如果在这里没有配置页面访问路径,页面被访问时就会报错;如果在这里定义了页面访问路径,框架页面文件中就会建立相应名称的文件夹及文件,不需要再手动添加文件夹和文件了,如图 3.4 所示。

(2) window: 配置窗口表现。它用于配置小程序的状态栏、导航条、标题、窗口背景色,可以设置导航条背景色(navigationBarBackgroundColor)、导航条文字(navigationBarTitleText)及导航条文字颜色(navigationBarTextStyle);还可以设置窗口是否可以下拉刷新(enablePullDownRefresh)(默认为不可以下拉刷新),设置窗口的背景色(backgroundColor)和下拉背景字体或者 loading 样式(backgroundTextStyle),如图 3.5 所示。

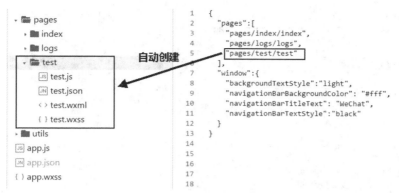

图 3.4 自动创建页面

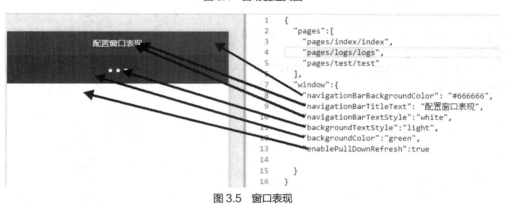

图 3.5 窗口表现

（3）tabBar：配置标签导航。标签导航是很多 App 都会采用的导航方式，微信小程序同样可以实现这样的效果，如图 3.6 所示。

图 3.6 莫凡商城 App 标签导航

怎么制作标签导航呢？需要在 app.json 文件里配置 tabBar 属性。tabBar 是一个对象，它可以配置标签导航文字的默认颜色、选中颜色，标签导航背景色及上边框颜色。上边框颜色目前可以配置为黑和白（black/white）两种。标签导航存放在 list 数组中，list 中的每个对象，对应一个标签导航，每个对象里可以配置标签导航的路径、导航名称、默认图标及选中图标，代码如下。

```
"tabBar": {
    "selectedColor": "#009966",          //选中时文字颜色
    "backgroundColor": "#ffffff",         //标签导航背景色
    "borderStyle": "white",               //标签导航标签颜色
    "color": "#999999",                   //文字默认颜色
    "list": [
        {
            "pagePath": "pages/index/index",                    //标签导航页面路径
            "text": "首页",                                      //标签导航文字
            "iconPath": "pages/images/bar/home-1.png",          //默认时图标
            "selectedIconPath": "pages/images/bar/home-0.png"   //选中时图标
        },
        {
            "pagePath": "pages/category/category",
            "text": "分类",
```

```
        "iconPath": "pages/images/bar/category-1.png",
        "selectedIconPath": "pages/images/bar/category-0.png"
      },
      {
        "pagePath": "pages/shoppingcart/shoppingcart",
        "text": "购物车",
        "iconPath": "pages/images/bar/cart-1.png",
        "selectedIconPath": "pages/images/bar/cart-0.png"
      },
      {
        "pagePath": "pages/me/me",
        "text": "我的",
        "iconPath": "pages/images/bar/me-1.png",
        "selectedIconPath": "pages/images/bar/me-0.png"
      }
    ]
  }
```

（4）networkTimeout：配置网络超时。它用于配置网络请求、文件上传、文件下载时最大的请求时间，超过这个时间，将不再请求。

（5）debug：配置 debug 模式，以方便微信小程序开发者调试开发程序。图 3.7 所示是开启了 debug 模式和没有开启 debug 模式的调试信息对比。

（a）没有开启 debug 模式

（b）开启了 debug 模式

图 3.7　debug 模式对比

从图 3.7 中可以看出，开启 debug 模式，可以看到每一步的调用情况、访问路径及错误信息，这样更加方便开发者进行调试工作。

（6）functionalPages：设置是否启用插件功能页，默认是关闭的。插件所有者小程序（指的是与插件 AppID 相同的小程序）需要设置这一项来启用插件功能页。

（7）subpackages：进行分包结构配置，启用分包加载时，声明项目分包结构。

（8）workers：使用 Worker 处理多线程任务时，设置 Worker 代码放置的目录。

（9）requiredBackgroundModes：声明需要后台运行的能力，类型为数组，目前支持 audio 后台音乐播放。

（10）plugins：声明小程序需要使用的插件。

（11）preloadRule：声明分包预下载的规则。

（12）resizable：在 iPad 上运行的小程序可以设置支持屏幕旋转。

（13）navigateToMiniProgramAppIdList：声明需要跳转的小程序列表。当小程序需要使用 wx.navigateToMiniProgram 接口跳转到其他小程序时，需要先在配置文件中声明需要跳转的小程序 AppId 列表，最多允许填写 10 个。

（14）usingComponents：进行全局自定义组件配置。在此处声明的自定义组件将被视为全局自定义组件，在小程序内的页面或自定义组件中可以直接使用而无需再声明。

（15）permission：进行小程序接口权限的相关设置，字段类型为 Object。

（16）sitemapLocation：指明 sitemap.json 文件的位置，默认为"sitemap.json"，即在 app.json 文件同级目录下。

app.json 文件作为全局配置文件，负责配置页面文件路径、窗口的表现、底部标签导航、网络连接超时、debug 模式等，配置起来比较容易。

3. app.wxss 小程序公共样式表文件

app.wxss 文件对 CSS 样式进行了扩展，它和 CSS 的使用方式一样，使用类选择器和行内样式的写法，兼容大部分 CSS 样式（有一些 CSS 样式不起作用），同时它还扩展了 CSS，形成了具有自己风格的样式文件。app.wxss 文件是对所有页面定义全局样式，如图 3.8 所示。只要页面有全局样式里的 class，就可以渲染全局样式里的效果，但如果页面又重新定义了这个 class 样式，就会把全局样式覆盖掉，使用自己的样式。

除了 app.wxss 文件提供的默认全局样式外，用户自己也可以定义一些全局样式，这样方便每个页面的使用，又不用在每个页面都写一次，达到一次定义，其他页面直接引用的复用效果。

```
/**app.wxss**/
.container {
  height: 100%;
  display: flex;
  flex-direction: column;
  align-items: center;
  justify-content: space-between;
  padding: 200rpx 0;
  box-sizing: border-box;
}
```

图 3.8 小程序公共样式表

4. project.config.json 小程序项目个性化配置文件

在使用微信小程序开发者工具时，开发者都会根据自己的喜好做一些个性化配置，如界面颜色、编译配置等，当换了另外一台计算机重新安装工具的时候，还要重新配置。基于这个考虑，小程序开发者工具在每个项目的根目录中都会生成一个 project.config.json 文件，开发者在工具上做的任何配置都会写入这个文件，当重新安装工具或者换计算机工作时，只要载入同一个项目的代码包，开发者工具就自动会恢复到当时开发项目时的个性化配置，其中包括编辑器的颜色、代码上传时自动压缩等一系列选项。

5. sitemap.json 小程序及其页面是否允许被微信索引的文件

小程序根目录下的 sitemap.json 文件用于配置小程序及其页面是否允许被微信索引，文件内容为一个 JSON 对象，如果没有 sitemap.json 文件，则默认为所有页面都允许被索引。配置示例如下所示。

```
{
  "rules":[{
    "action": "allow",
    "page": "path/to/page",
    "params": ["a",  "b"],
    "matching": "exact"
  }, {
    "action": "disallow",
    "page": "path/to/page"
  }]
}
```

（1）path/to/page?a=1&b=2 这个页面路径会被优先索引。
（2）path/to/page 这个页面路径不会被索引。
（3）path/to/page?a=1 这个页面路径不会被索引。
（4）path/to/page?a=1&b=2&c=3 这个页面路径不会被索引。
（5）其他页面都会被索引。
matching 匹配规则说明如下。
（1）exact：当小程序页面的参数列表等于 params 时，规则命中。
（2）inclusive：当小程序页面的参数列表包含 params 时，规则命中。
（3）exclusive：当小程序页面的参数列表与 params 的交集为空时，规则命中。
（4）partial：当小程序页面的参数列表与 params 的交集不为空时，规则命中。

3.1.2 项目实战：任务 1——实现底部标签导航功能

1. 任务目标

通过实现莫凡商城底部标签导航功能，学会设计小程序的底部标签导航及窗口显示。

莫凡商城底部标签导航分为"首页""分类""购物车""我的"4 个页面。标签导航选中时文字和图标均呈现为绿色；默认文字和图标均为灰色，如图 3.9 所示。

图 3.9 底部标签导航

2. 任务实施

（1）在 app.json 文件的 pages 对象里配置"首页""分类""购物车""我的"页面路径，代码如下。

```
"pages": [
    "pages/index/index",
    "pages/category/category",
    "pages/shoppingcart/shoppingcart",
    "pages/me/me"
]
```

（2）在 window 对象里配置窗口文字为"莫凡商城"，背景颜色为绿色，代码如下。

```
"window": {
    "backgroundTextStyle": "light",
    "navigationBarBackgroundColor": "#009966",
    "navigationBarTitleText": "莫凡商城",
    "navigationBarTextStyle": "white"
}
```

（3）在 tabBar 标签导航里配置导航文字和图标，代码如下。

```
"tabBar": {
    "selectedColor": "#009966",
    "backgroundColor": "#ffffff",
    "borderStyle": "black",
    "color": "#999999",
    "list": [
        {
            "pagePath": "pages/index/index",
            "text": "首页",
            "iconPath": "pages/images/bar/home-1.png",
            "selectedIconPath": "pages/images/bar/home-0.png"
        },
        {
            "pagePath": "pages/category/category",
            "text": "分类",
            "iconPath": "pages/images/bar/category-1.png",
            "selectedIconPath": "pages/images/bar/category-0.png"
        },
        {
            "pagePath": "pages/shoppingcart/shoppingcart",
            "text": "购物车",
            "iconPath": "pages/images/bar/cart-1.png",
            "selectedIconPath": "pages/images/bar/cart-0.png"
        },
        {
            "pagePath": "pages/me/me",
            "text": "我的",
            "iconPath": "pages/images/bar/me-1.png",
            "selectedIconPath": "pages/images/bar/me-0.png"
        }
    ]
}
```

通过这 3 个步骤的配置，就可以实现莫凡商城底部标签导航的配置了，效果如图 3.10 所示。

在设计底部标签导航的时候，需要在 app.json 文件里针对 tabBar 对象进行配置，可以配置导航的标题、导航的标题图、选中的文字颜色和默认的文字颜色。

图 3.10　莫凡商城小程序底部标签导航

3.1.3　工具类文件

在微信小程序框架目录里有一个 utils 文件夹，它用来存放工具栏的 js 函数，如可以放置一些日期格式化、时间格式化的常用函数，定义完这些函数后，通过 module.exports 将定义的函数名称注册进来，在其他页面才可以使用这些函数，图 3.11 所示为时间格式化工具类。

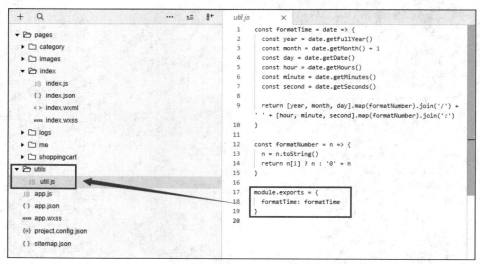

图 3.11　utils.js 工具类文件

3.1.4　框架页面文件

一个小程序的框架页面文件由 5 个文件组成，分别是页面逻辑文件（JS）、页面配置文件（JSON）、页面结构文件（WXML）、小程序脚本语言文件（WXS）、页面样式表文件（WXSS），如表 3.3 所示。

表 3.3　框架页面文件

文件类型	是否必填	作用
JS	是	编写页面逻辑
JSON	否	编写页面配置
WXML	是	编写页面结构
WXS	否	编写小程序脚本语言
WXSS	否	编写页面样式表

微信小程序的框架页面文件，都是放置在 pages 文件夹下的，如图 3.12 所示。

每个页面都有一个独立的文件夹，例如，日志页面 logs 文件夹下放置了 5 个文件，logs.js 文件进行业务路径处理；logs.json 文件进行页面的配置，可以覆盖全局 app.json 文件配置；logs.wxml 文件是页面结构文件，负责渲染页面；logs.wxs 文件是小程序脚本语言文件，在创建页面的时候不会自动生成该文件，需要使用的时候创建即可；logs.wxss 文件是针对 logs.wxml 文件页面的样式文件。

注意：WXS（WeiXin Script）是小程序的一套脚本语言，结合 WXML（WeiXin Markup Language），可以构建出页面的结构。

图 3.12　页面文件

3.2　微信小程序逻辑层框架接口

要让小程序顺利运行，首先要注册小程序和其中的页面，并为它们设置生命周期。在 3.1.1 小节中，我们介绍了框架全局文件，其中，app.js 文件是小程序的逻辑文件，我们在这里定义全局数据和函数，本节详细讲解在 app.js 文件中注册小程序和页面的方法。

3.2.1　使用 App()函数注册小程序

在 app.js 小程序逻辑文件里，App（Object object）函数是用来注册小程序的，有了这个函数，才能说明这个项目是小程序项目。App()函数有一个 Object 参数，在注册小程序的同时可以指定小程序的生命周期函数。

App()函数必须在 app.js 文件中调用，必须调用且只能调用一次，否则会出现无法预期的后果。示例代码如下：

```
App({
  onLaunch (options) {
    // Do something initial when launch.
  },
  onShow (options) {
    // Do something when show.
  },
  onHide () {
    // Do something when hide.
  },
  onError (msg) {
    console.log(msg)
  },
  onPageNotFound(msg){
    console.log(msg)
  },
  globalData: 'I am global data'
})
```

App()中的函数详解如下。

（1）onLaunch：生命周期回调函数，监听小程序初始化，在小程序初始化完成时触发，全局只触发一次。

（2）onShow：生命周期回调函数，监听小程序启动或切到前台，在小程序启动，或从后台进入前台显示时触发。

（3）onHide：生命周期回调函数，监听小程序切到后台，在小程序从前台进入后台时触发。

（4）onError：错误监听函数，在小程序发生脚本错误或 API（应用程序接口）调用报错时触发。

（5）onPageNotFound：页面不存在监听函数，在小程序要打开的页面不存在时触发。

我们可以通过 AppObject getApp（Object object）函数获取小程序全局唯一的 App 实例，然后就可以调用 app.js 文件里配置的自定义函数和自定义全局数据了，代码如下。

```
var app = getApp()
console.log(appInstance.globalData.host) //获取在 globalData 全局数据里配置的域名
```

getApp()函数在页面的.js 文件里执行就可以获取到小程序全局唯一的 App 实例，莫凡商城小程序也是采用这样的方式来获取在 globalData 全局数据中配置的域名的。

注意：

（1）不要在定义于 App()内的函数中调用 getApp()，使用 this 就可以拿到 App 实例。

（2）通过 getApp()获取实例之后，不要私自调用生命周期函数。

3.2.2 使用 Page()函数注册页面

在 app.js 文件里使用 App()函数可以注册小程序，在框架页面文件里，根据配置的路径，会生成*.js、*.wxml、*.json、*.wxml 4 个文件（*.wxs 文件不会自动生成，需要时才手动创建），我们在*.js 文件里需要使用 Page()函数来注册一个页面，指定页面的初始数据、生命周期回调、事件处理函数等，代码如下。

```js
//index.js
Page({
  data: {
    text: "This is page data."
  },
  onLoad: function(options) {
    // Do some initialize when page load.
  },
  onReady: function() {
    // Do something when page ready.
  },
  onShow: function() {
    // Do something when page show.
  },
  onHide: function() {
    // Do something when page hide.
  },
  onUnload: function() {
    // Do something when page close.
  },
  onPullDownRefresh: function() {
    // Do something when pull down.
  },
  onReachBottom: function() {
    // Do something when page reach bottom.
  },
  onShareAppMessage: function () {
    // return custom share data when user share.
  },
```

```
onPageScroll: function() {
  // Do something when page scroll
},
onResize: function() {
  // Do something when page resize
},
onTabItemTap(item) {
  console.log(item.index)
  console.log(item.pagePath)
  console.log(item.text)
},
// Event handler.
viewTap: function() {
  this.setData({
    text: 'Set some data for updating view.'
  }, function() {
    // this is setData callback
  })
},
customData: {
  hi: 'MINA'
}
})
```

（1）data 对象：data 是页面第一次渲染时使用的初始数据，页面加载时，data 将会以 JSON 字符串的形式由逻辑层传至渲染层，因此 data 中的数据必须是可以转换成 JSON 的类型：字符串、数字、布尔值、对象或数组。

JSON（JavaScript Object Notation）是一种轻量级的数据交换格式，示例如下所示。

```
{
  "employees": [
    {
      "firstName": "Bill",
      "lastName": "Gates"
    }
  ]
}
```

（2）onLoad（Object query）：生命周期回调函数，页面加载时触发。一个页面只会调用一次，可以在 onLoad 的参数中获取打开当前页面路径中的参数。

（3）onReady()：生命周期回调函数，页面初次渲染完成时触发。一个页面只会调用一次，代表页面已经准备妥当，可以和视图层进行交互。

（4）onShow()：生命周期回调函数，页面显示/切入前台时触发。

（5）onHide()：生命周期回调函数，页面隐藏/切入后台时触发，当调用 navigateTo 或底部导航标签切换时调用。

（6）onUnload()：生命周期回调函数，页面卸载时触发，当调用 redirectTo 或 navigateBack 的时候调用。

（7）onPullDownRefresh()：页面事件处理函数，监听用户下拉刷新事件，需要在 app.json 文件的 window 选项中或页面配置中开启 enablePullDownRefresh。可以通过 wx.startPullDownRefresh 触发下拉刷新，调用后触发下拉刷新动画，效果与用户手动下拉刷新一致。当处理完数据刷新后，wx.stopPullDownRefresh 可以停止当前页面的下拉刷新。

（8）onReachBottom()：页面事件处理函数，监听用户上拉触底事件，可以在 app.json 文件的 window 选项中或页面配置中设置触发距离 onReachBottomDistance。在触发距离内滑动期间，本事件只会被触发一次。

（9）onShareAppMessage（Object object）：页面事件处理函数，监听用户单击页面内转发按钮（button 组件 open-type="share"）或右上角菜单中的"转发"按钮的行为，并自定义转发内容，只有定义了此事件处理函数，右上角菜单中才会显示"转发"按钮。

（10）onPageScroll（Object object）：页面事件处理函数，监听用户滑动页面事件。

（11）onResize()：页面事件处理函数，监听页面尺寸发生改变的事件。对于自定义组件，可以使用 resize 生命周期来监听。回调函数中将返回显示区域的尺寸信息。

（12）onTabItemTap()：页面事件处理函数，单击 tab（导航标签）时触发，可以获取 tab 的序号、页面路径、按钮文字。

（13）组件事件处理函数：Page 中还可以定义组件事件处理函数。在渲染层的组件中加入事件绑定，当事件被触发时，就会执行 Page 中定义的事件处理函数，如"注册"按钮可以自定义绑定 register() 事件、"登录"按钮可以自定义绑定 login()等。

注意：通过 getCurrentPages()函数可以获取当前页面栈，返回页面数组对象 PageObject[]。数组中第一个元素为首页，最后一个元素为当前页面，不要尝试修改页面栈，会导致路由及页面状态错误。不要在小程序初始化的时候调用 getCurrentPages()，此时 page 还没有生成。

3.3 微信小程序 WXML 视图层

完成小程序和页面的注册，我们就可以开始构建小程序页面视图了。WXML（WeiXin Markup Language）是视图层框架设计的一套标签语言，结合基础组件、事件系统，可以构建出页面的视图。

3.3.1 WXML 标签语言

在框架页面文件的*.wxml 文件里，可以利用 WXML 标签语言来构建小程序页面视图，构建视图页面内容就需要用到组件，如在页面里想显示出"你好，微信小程序"，代码如下。

```
<view> 你好，微信小程序</view>
```

3.3.2 动态绑定数据

在*.wxml 文件里，可以使用 view 组件进行布局设计（将在 4.2 节中详细讲解）。但在 3.3.1 小节中，显示的内容是直接写在 view 组件中的，不是动态数据，如何动态地绑定数据呢？

.wxml 文件中使用的动态数据，都来自于.js 文件中 Page()函数的 data 对象。动态数据绑定就是在*.wxml 文件中，通过双大括号（{{}}）将*.js 文件中定义的变量包起来，放在 view 组件中，这样就可以实现数据动态绑定效果了。

示例代码如下。

```
<!--index.wxml-->
<view> {{ message }} </view>

//index.js
Page({
  data: {
    message: '你好，微信小程序'
  }
})
```

慕课视频

微信小程序 WXML 视图层

3.3.3 组件属性动态绑定数据

组件属性动态绑定数据，是将*.js 文件 data 对象里的数据绑定到小程序的组件上，示例代码如下。

```
<!--index.wxml-->
<view id="{{id}}"> </view>
```

```
//index.js
Page({
  data: {
    id: 0
  }
})
```

3.3.4 控制属性动态绑定数据

控制属性动态绑定数据是通过条件判断 if 语句来控制的,如果满足条件判断,则执行该语句,否则不执行,示例代码如下。

```
<!--index.wxml-->
<view wx:if="{{flag}}"> </view>
```

```
//index.js
Page({
  data: {
    flag: true
  }
})
```

3.3.5 关键字动态绑定数据

关键字动态绑定数据用于将组件的一些关键字绑定数据,如复选框组件,checked 关键字如果等于 true,则代表选中复选框,等于 false 则代表不选中复选框,示例代码如下。

```
<checkbox checked="{{false}}"> </checkbox>
```

不要直接写 checked="false",否则其计算结果是一个字符串,转换成 boolean 类型后代表真值。

3.3.6 运算

可以在 {{}} 内进行简单的运算,小程序支持以下几种方式的运算。

(1)三元运算,示例代码如下。

```
<view hidden="{{flag ? true : false}}"> Hidden </view>
```

(2)数学运算,示例代码如下。

```
<!--index.wxml-->
<view> {{a + b}} + {{c}} + d </view>
```

```
//index.js
Page({
  data: {
    a: 1,
    b: 2,
    c: 3
  }
})
```

view 中的内容为 3 + 3 + d。

(3)逻辑判断,示例代码如下。

```
<view wx:if="{{length > 5}}"> </view>
```

(4)字符串运算,示例代码如下。

```
<!--index.wxml-->
<view>{{"hello" + name}}</view>
```

```
//index.js
Page({
```

```
    data:{
      name: 'MINA'
    }
})
```
（5）数据路径运算，示例代码如下。
```
<!--index.wxml-->
<view>{{object.key}} {{array[0]}}</view>
```
```
//index.js
Page({
  data: {
    object: {
      key: 'Hello '
    },
    array: ['MINA']
  }
})
```

3.4 微信小程序 WXSS 样式渲染

WXSS（WeiXin Style Sheets）是一套样式语言，用于描述 WXML 的组件样式。WXSS 具有 CSS（Cascading Style Sheets，层叠样式表）的大部分特性，同时 WXSS 对 CSS 进行了扩充及修改，用来决定 WXML 组件的显示效果。与 CSS 相比，WXSS 在尺寸单位和样式导入上进行了扩展。

3.4.1 尺寸单位

WXSS 的尺寸单位是 rpx（responsive pixel），它可以根据屏幕宽度进行自适应。屏幕宽度规定为 750 rpx。例如，iPhone6 的屏幕宽度为 375 px，共有 750 个物理像素，则 750 rpx = 375 px = 750 个物理像素，1 rpx = 0.5 px = 1 个物理像素。rpx 与 px 的换算关系如表 3.4 所示。

表 3.4 rpx 与 px 的换算

设备	rpx 换算成 px（屏幕宽度/750）	px 换算成 rpx（750/屏幕宽度）
iPhone5	1 rpx = 0.42 px	1 px = 2.34 rpx
iPhone6	1 rpx = 0.5 px	1 px = 2 rpx
iPhone6 Plus	1 rpx = 0.552 px	1 px = 1.81 rpx

3.4.2 样式导入

使用@import 语句可以导入外联样式表，@import 后跟需要导入的外联样式表的相对路径，用;表示语句结束，示例代码如下。
```
/** common.wxss **/
.small-p {
  padding:5px;
}
```
```
/** app.wxss **/
@import "common.wxss";
.middle-p {
  padding:15px;
}
```
这样，在 app.wxss 文件里，可以将 common.wxss 文件样式引入进来使用。

定义在 app.wxss 文件中的样式为全局样式，作用于每一个页面。在每个页面的*.wxss 文件中定义的样式为局部样式，只作用于对应的页面，并会覆盖 app.wxss 文件中相同的选择器。

3.4.3 内联样式

在 WXML 视图组件中，可以使用 style、class 属性来控制组件的样式。

（1）style：用于接收动态的样式，在运行时会进行解析，静态的样式统一写到 class 中，要尽量避免将静态的样式写进 style 中，以免影响渲染速度，style 属性示例代码如下。

```
<view style="color:red;" /> //静态的样式写进 style，尽量避免使用
<view style="color:{{color}}" /> //动态获取
```

（2）class：用于指定样式规则，其属性值是样式规则中类选择器名（样式类名）的集合，样式类名不需要带上"."，样式类名之间用空格分隔。

```
<view class="normal_view" />
```

3.4.4 选择器

WXSS 样式渲染支持用选择器来控制，现在支持的选择器如表 3.5 所示。

表 3.5 支持的选择器

选择器	样例	样例描述
.class	.intro	选择所有拥有 class="intro"的组件
#id	#firstname	选择拥有 id="firstname"的组件
element	view	选择所有 view 组件
element, element	view, checkbox	选择所有文档的 view 组件和所有的 checkbox 组件
::after	view::after	在 view 组件后边插入内容
::before	view::before	在 view 组件前边插入内容

3.4.5 常用样式属性

常用样式包括 display（显示）、position（定位）、float（浮动）、background（背景）、border（边框）、outline（轮廓）、text（文本）、font（字体）、margin（外边距）、padding（填充）等。

（1）display（显示）样式，其属性和说明如表 3.6 所示。

表 3.6 display 样式的属性和说明

属性	说明
flex	多栏多列布局，常和 flex-direction: row/column 一起使用
inline-block	行内块元素
inline	此元素会被显示为内联元素，元素前后没有换行符
inline-table	作为内联表格来显示（类似<table>），表格前后没有换行符
inline-flex	将对象作为内联块级弹性伸缩盒显示
none	此元素不会被显示
lock	此元素将显示为块级元素，此元素前后会带有换行符
list-item	此元素会作为列表显示
table	会作为块级表格来显示（类似<table>），表格前后带有换行符
table-caption	作为一个表格标题显示（类似<caption>）
table-cell	作为一个表格单元格显示（类似<td>和<th>）
table-column	作为一个单元格列显示（类似<col>）
table-column-group	作为一个或多个列的分组来显示（类似<colgroup>）
table-row	作为一个表格行显示（类似<tr>）

属性	说明
table-row-group	作为一个或多个行的分组来显示（类似<tbody>）
table-header-group	作为一个或多个行的分组来显示（类似<thead>）
table-footer-group	作为一个或多个行的分组来显示（类似<tfoot>）
inherit	从父元素继承display属性的值

（2）position（定位）样式，其属性和说明如表3.7所示。

表3.7 position样式的属性和说明

属性	说明
absolute	生成绝对定位的元素，相对于static定位以外的第一个父元素进行定位。元素的位置通过left、top、right及bottom属性进行规定
relative	生成相对定位的元素，相对于其正常位置进行定位。因此，"left: 20"会向元素的LEFT位置添加20像素
fixed	生成绝对定位的元素，相对于浏览器窗口进行定位。元素的位置通过left、top、right及bottom属性进行规定
static	默认值，没有定位，元素出现在正常的流中（忽略top、bottom、left、right或者z-index声明）
inline-flex	将对象作为内联块级弹性伸缩盒显示
inherit	规定应该从父元素继承position属性的值

（3）float（浮动）样式，其属性和说明如表3.8所示。

表3.8 float样式的属性和说明

属性	说明
left	元素向左浮动
right	元素向右浮动
none	默认值，元素不浮动，并会显示其在文本中出现的位置
inherit	规定应该从父元素继承float属性的值

（4）background（背景）样式，其属性和说明如表3.9所示。

表3.9 background样式的属性和说明

属性	说明
background	简写属性，作用是将背景属性设置在一个声明中 background: color position size repeat origin clip attachment image;
background-color	指定要使用的背景颜色
background-position	指定背景图像的位置 background-position: center
background-size	指定背景图像的大小 background-size: 80px 60px;（宽度、高度）
background-repeat	指定如何重复背景图像 repeat, repeat-x, repeat-y, no-repeat, inherit
background-origin	指定背景图像的定位区域 padding-box。背景图像填充框的相对位置
border-box	背景图像边界框的相对位置
content-box	背景图像相对位置的内容框
background-clip	指定背景图像的绘画区域。属性值，同上
background-attachment	设置背景图像是否固定或者随着页面的其余部分滚动
scroll	背景图像随页面的其余部分滚动。这是默认属性
fixed	背景图像是固定的
inherit	指定background-attachment的设置应该从父元素继承

续表

属性	说明
local	背景图像随滚动元素滚动
background-image	指定要使用的一个或多个背景图像，使用 url ('URL') 提供图像的 URL

（5）border（边框）样式，其属性和说明如表 3.10 所示。

表 3.10 border 样式的属性和说明

属性	说明	属性值
border	简写属性，用于把针对 4 个边的属性设置在一个声明中	border: 5px solid red;
border-width	用于为元素的所有边框设置宽度，或者单独地为各边框设置宽度	thin、medium、thick、length
border-style	设置元素所有边框的样式，或者单独地为各边设置边框样式	solid、dashed、dotted、double 等
border-color	设置元素的所有边框中可见部分的颜色，或为 4 个边分别设置颜色	border-color: red;

（6）outline（轮廓）样式，其属性和说明如表 3.11 所示。

表 3.11 outline 样式的属性和说明

属性	说明	属性值
outline	在一个声明中设置所有的外边框属性	outline: outline-color, outline-style, outline-width
outline-color	设置外边框的颜色	
outline-style	设置外边框的样式	Solid、dashed、dotted、double 等
outline-width	设置外边框的宽度	thin medium thick length

（7）text（文本）样式，其属性和说明如表 3.12 所示。

表 3.12 text 样式的属性和说明

属性	说明	属性值
color	设置文本颜色	
direction	设置文本方向	ltr：文本方向从左到右 rtl：文本方向从右到左
letter-spacing	设置字符间距	
line-height	设置行高	
text-align	对齐元素中的文本	left：把文本排列到左边。默认值，由浏览器决定 right：把文本排列到右边 center：把文本排列到中间 justify：实现两端对齐文本效果 inherit：规定应该从父元素继承 text-align 属性的值
text-decoration	向文本添加修饰	underline：定义文本下的一条线 overline：定义文本上的一条线 line-through：定义穿过文本下的一条线 blink：定义闪烁的文本
text-indent	缩进元素中文本的首行	
text-shadow	设置文本阴影	text-shadow: h-shadow v-shadow blur color; h-shadow：水平阴影的位置，允许负值 v-shadow：垂直阴影的位置，允许负值 blur：模糊的距离 color：阴影的颜色

续表

属性	说明	属性值
text-transform	控制元素中的字母	capitalize：文本中的每个单词都以大写字母开头 uppercase：定义仅有大写字母 lowercase：定义无大写字母，仅有小写字母
vertical-align	设置元素的垂直对齐	baseline、sub、super、top、text-top、middle、bottom、text-bottom、length、%、inherit
white-space	设置元素中空白的处理方式	normal、pre、nowrap、pre-wrap、pre-line、inherit
word-spacing	设置字间距	normal、length、inherit

（8）font（字体样式），其属性和说明如表3.13所示。

表3.13 font样式的属性和说明

属性	说明	属性值
font	在一个声明中设置所有字体属性	font: font-style font-variant font-weight font-size/line-height font-family（按顺序）
font-style	指定文本的字体样式	normal：默认值。浏览器会显示一个标准的字体样式 italic：浏览器会显示一个斜体的字体样式 oblique：浏览器会显示一个倾斜的字体样式 inherit：规定应该从父元素继承字体样式
font-variant	以小型大写字体或者正常字体显示文本	normal：默认值。浏览器会显示一个标准的字体 small-caps：浏览器会显示小型大写字母的字体 inherit：规定应该从父元素继承font-variant属性的值
font-weight	指定字体的粗细	normal：默认值。定义标准的字符 bold：定义粗体字符 bolder：定义更粗的字符 lighter：定义更细的字符 inherit：规定应该从父元素继承字体的粗细
font-size	指定文本的字体大小	smaller：把font-size设置为比父元素更小的尺寸 larger：把font-size设置为比父元素更大的尺寸 length：把font-size设置为一个固定的值 %：把font-size设置为基于父元素的一个百分比值
font-family	指定文本的字体系列	

（9）margin（外边距）样式，其属性和说明如表3.14所示。

表3.14 margin样式的属性和说明

属性	说明	属性值
margin	在一个声明中设置所有外边距属性	margin：10px 5px 15px 20px;（上边距，右边距，下边距，左边距）
margin-top	设置元素的上外边距	
margin-right	设置元素的右外边距	
margin-bottom	设置元素的下外边距	
margin-left	设置元素的左外边距	

（10）padding（填充）样式，其属性和说明如表3.15所示。

表3.15 padding样式的属性和说明

属性	说明	属性值
padding	使用缩写属性设置在一个声明中的所有填充属性	padding：10px 5px 15px 20px;（上填充，右填充，下填充，左填充）

续表

属性	说明	属性值
padding-top	设置元素的顶部填充	
padding-right	设置元素的右部填充	
padding-bottom	设置元素的底部填充	
padding-left	设置元素的左部填充	

3.5 微信小程序条件渲染

在编写微信小程序时,经常需要进行条件判断,以确定是否需要渲染某代码块。

3.5.1 使用 wx: if 判断单个组件

在微信小程序框架里,使用 wx: if="{{condition}}" 来进行条件判断,判断是否需要渲染该代码块,示例代码如下。

```
<view wx:if="{{condition}}"> 你好,欢迎学习微信小程序 </view>
```

使用 wx: elif 和 wx: else 来添加一个 else 代码块。

```
<view wx:if="{{length > 5}}"> 长度大于 5 </view>
<view wx:elif="{{length > 2}}"> 长度大于 2 </view>
<view wx:else> 长度为 3 </view>
```

```
//index.js
Page({
  data: {
    length: 8
  }
})
```

当 length=8 时,输出结果为长度大于 5,执行第一个判断条件。

3.5.2 使用 block wx: if 判断多个组件

因为 wx: if 是一个控制属性,需要将它添加到一个组件上。但是如果我们想一次性判断多个组件,可以使用一个 <block/> 标签将多个组件包装起来,并在上边使用 wx: if 控制属性,示例代码如下。

```
<block wx:if="{{true}}">
  <view> 内容 1</view>
  <view> 内容 2</view>
</block>
```

<block/> 并不是一个组件,它仅仅是一个包装元素,不会在页面中做任何渲染,只接受控制属性。

3.6 微信小程序列表渲染

我们经常需要将一些内容以列表的形式显示出来,这就要用到微信小程序的列表渲染。如果只是将数据以列表的形式显示出来,那么直接一行行显示就行。但如果数据的显示是动态的,这种方式就不能解决问题了。微信小程序提供了 wx: for 的方式来解决这个问题。

3.6.1 使用 wx: for 列表渲染单个组件

在组件上使用 wx: for 控制属性绑定一个数组,即可使用数组中各项的数据重复渲染该组件。默认数组当前项的下标变量名为 index,数组当前项的变量名为 item,示例代码如下。

```
<view wx:for="{{array}}">
  {{index}}: {{item.name}}
</view>
```

```
//index.js
Page({
  data: {
    array: [{
      name: 'tom',
    }, {
      name: 'kevin'
    }]
  }
})
```

使用 wx: for-item 可以指定数组当前元素的变量名，使用 wx: for-index 可以指定数组当前下标的变量名，示例代码如下。

```
<view wx:for="{{array}}" wx:for-index="idx" wx:for-item="itemName">
  {{idx}}: {{itemName.name}}
</view>
```

3.6.2 使用 block wx: for 列表渲染多个组件

wx: for 应用在某一个组件上，但是当想渲染一个包含多节点的结构块时，wx: for 就需要应用在 `<block/>` 标签上，示例代码如下。

```
<block wx:for="{{[1, 2, 3]}}">
  <view> {{index}}: </view>
  <view> {{item}} </view>
</block>
```

3.6.3 使用 wx: key 指定唯一标识符

如果列表中项目的位置会动态改变，或者有新的项目添加到列表中，并且希望列表中的项目保持自己的特征和状态（如 `<input/>` 中的输入内容，`<switch/>` 的选中状态），就需要使用 wx: key 来指定列表中项目的唯一标识符。

wx: key 的值以下列两种形式提供。

（1）字符串：代表在 for 循环的集合中值的某个属性，该属性的值需要是列表中唯一的字符串或数字，且不能动态改变。

（2）保留关键字：*this 代表在 for 循环中的 item 本身，这种表示需要 item 本身是一个唯一的字符串或者数字，当数据改变触发渲染层重新渲染的时候，会校正带有 key 的组件，框架会确保它们被重新排序，而不是重新创建，以确保组件保持自身的状态，并且提高列表渲染时的效率。

示例代码如下。

```
<switch wx:for="{{objectArray}}" wx:key="unique" style="display: block;"> {{item.id}} </switch>
Page({
  data: {
    objectArray: [
      {id: 5, unique: 'unique_5'},
      {id: 4, unique: 'unique_4'},
      {id: 3, unique: 'unique_3'},
      {id: 2, unique: 'unique_2'},
      {id: 1, unique: 'unique_1'},
      {id: 0, unique: 'unique_0'},
    ]
  }
})
```

注意：如不提供 wx: key，会报一个 warning，如果明确知道该列表是静态，或者不必关注其顺序，可以选择忽略。

3.7 项目实战：任务 4——实现"我的"界面列表式导航功能（1）

1. 任务目标

通过实现"我的"界面列表式导航功能，学会使用 WXML 标签语言进行页面布局，使用 WXSS 进行样式渲染，进行数据绑定及列表式导航功能的应用。

"我的"界面可以分为 3 部分：第 1 部分是登录相关的内容，包括头像和昵称；第 2 部分是订单相关的内容，包括待付款、待收货、已完成的图标和文字；第 3 部分采用列表式导航，包括我的消息、我的收藏、账户余额、修改密码、意见反馈、清除缓存和知识扩展，如图 3.13 所示。

慕课视频

项目实战：实现"我的"界面列表式导航

图 3.13 "我的"界面

2. 任务实施

（1）在 me.wxml 文件里进行页面布局设计，代码如下。

```
<view class="content">
  <!-- 登录相关-->
  <view class="head">
    <view class="headIcon">
      <image src="/pages/images/icon/head.jpg" style="width:70px;height:70px;"></image>
    </view>
    <view class="login">
      <navigator url="../login/login" hover-class="navigator-hover">{{nickName}}</navigator>
    </view>
    <view class="detail">
      <text>></text>
    </view>
  </view>
  <view class="hr"></view>
<!--订单相关-->
<view style="display:flex;flex-direction:row;">
```

```
      <view class="order">我的订单</view>
      <view class="detail2">
        <text></text>
      </view>
    </view>
    <view class="line"></view>
    <view class="nav">
      <view class="nav-item" bindtap="nav" id="0" data-status='1'>
        <view>
          <image src="/pages/images/icon/dfk.png" style="width:28px;height:25px;"></image>
        </view>
        <view>待付款</view>
      </view>
      <view class="nav-item" bindtap="nav" id="1" data-status='3'>
        <view>
          <image src="/pages/images/icon/dsh.png" style="width:36px;height:27px;"></image>
        </view>
        <view>待收货</view>
      </view>
      <view class="nav-item" bindtap="nav" id="2" data-status='4'>
        <view>
          <image src="/pages/images/icon/dpj.png" style="width:31px;height:28px;"></image>
        </view>
        <view>已完成</view>
      </view>
    </view>
    <view class="hr"></view>
    <!--列表式导航相关-->
    <view class="item">
      <view class="order">我的消息</view>
      <view class="detail2">
        <text></text>
      </view>
    </view>
    <view class="line"></view>
    <view class="item">
      <view class="order">我的收藏</view>
      <view class="detail2">
        <text></text>
      </view>
    </view>
    <view class="line"></view>
    <view class="item">
      <view class="order">账户余额</view>
      <view class="detail2">
        <text>0.00 元 </text>
      </view>
    </view>
    <view class="line"></view>
    <view class="hr"></view>
    <view class="item" bindtap="updatePwd">
      <view class="order">修改密码</view>
      <view class="detail2">
        <text></text>
      </view>
    </view>
    <view class="line"></view>
    <view class="item" bindtap="opinion">
      <view class="order">意见反馈</view>
      <view class="detail2">
```

```
        <text>></text>
      </view>
    </view>
    <view class="line"></view>
    <view class="item" bindtap='clearStore'>
      <view class="order">清除缓存</view>
      <view class="detail2">
        <text >></text>
      </view>
    </view>
    <view class="line"></view>
    <view class="hr"></view>
    <view class="line"></view>
    <view class="item">
      <view class="order">知识扩展</view>
      <view class="detail2">
        <text >></text>
      </view>
    </view>
    <view class="hr"></view>
</view>
```

（2）在 me.wxss 文件里进行页面样式渲染，代码如下。

```
.head{
    width:100%;
    height: 90px;
    background-color: #009966;
    display: flex;
    flex-direction: row;
}
.headIcon{
    margin: 10px;
}
.headIcon image{
  border-radius:50%;
}
.login{
    color: #ffffff;
    font-size: 15px;
    font-weight: bold;
    position: absolute;
    left:100px;
    margin-top:30px;
}
.detail{
    color: #ffffff;
    font-size: 15px;
    position: absolute;
    right: 10px;
    margin-top: 30px;
}
.nav{
    display: flex;
    flex-direction: row;
    padding-top:10px;
    padding-bottom: 10px;
}
.nav-item{
    width: 25%;
    font-size: 13px;
```

```
        text-align: center;
        margin:0 auto;
}
.hr{
    width: 100%;
    height: 15px;
    background-color: #f5f5f5;
}
.order{
    padding-top:15px;
    padding-left: 15px;
    padding-bottom:15px;
    font-size:15px;
}
.detail2{
    font-size: 15px;
    position: absolute;
    right: 10px;
    margin-top:15px;
    color: #888888;
}
.line{
     height: 1px;
    width: 100%;
    background-color: #666666;
    opacity: 0.2;
}
.item{
  display:flex;
  flex-direction:row;
}
```

（3）在 me.js 文件里实现 nickName 昵称数据的动态绑定，代码如下。

```
Page({
  data: {
    nickName: '立即登录'
  }
})
```

（4）在 me.json 文件里修改页面导航标题，如果在各个页面的.json 文件里重新设置导航标题，就显示各个页面里设置的标题，否则显示 app.json 文件里配置的全局导航标题，代码如下。

```
{
    "navigationBarTitleText": "我的"
}
```

这样我们就完成了"我的"界面列表式导航设计，很多 App 应用也都会采用这样的导航，学会一种列表式导航设计，在设计其他应用的时候，就可以直接借鉴使用了。

3.8 小结

本章介绍了莫凡商城小程序的项目结构，详细介绍了各个项目结构的应用，包括底部标签导航设计、小程序逻辑层接口应用、WXML 视图层应用、WXSS 样式渲染、wx: if 条件判断、wx: for 列表渲染等内容，这些内容是最基础的，也是最应该掌握的，能够为后续的学习打下良好的基础。另外，通过完成任务 1——实现底部标签导航功能和任务 4——实现"我的"界面列表式导航功能，我们进一步学习了这些基础知识在实际开发中是如何应用的。

第4章
莫凡商城首页静态布局设计

莫凡商城首页包含 3 个部分的功能：第 1 部分是用来搜索莫凡商城商品的搜索区域；第 2 部分是通过海报轮播效果显示的轮播广告；第 3 部分用来展示图书商品，每个区域展示 3 本图书，可以通过单击"查看更多"按钮找到更多的图书商品，如图 4.1 所示。

图 4.1 首页效果

4.1 首页需求分析与知识点

莫凡商城首页包含搜索、海报轮播、展示图书商品、查看更多图书商品功能，它的布局设计需要用到视图容器组件，如 view 视图容器组件、swiper 滑块视图容器组件、text 文本组件、image 图片组件等。我们会详细介绍这些组件及其相关组件的使用方法，海报轮播效果使用 swiper 滑块视图容器组件来实现；单击搜索区域和查看更多图书商品时会进行页面跳转，会用到导航组件和导航 API；界面的样式需要用 WXSS 样式来渲染，进行页面美化布局，最终才能完成莫凡商城首页的静态布局设计。

4.2 视图容器组件在首页中的应用

视图容器组件是用来进行页面布局的,不同的视图容器组件可以用来实现不同的布局效果，view 视图容器组件是基本的容器组件，scroll-view 是可滚动视图容器组件，swiper 是滑块视图容器组件，movable-view 是可移动视图容器组件，cover-view 是覆盖原生组件的视图容器组件。

4.2.1 view 视图容器组件

view 视图容器组件是 WXML 界面布局的基础组件，也是最常用的界面布局组件，它的使用和 HTML 里的 DIV 功能类似。view 视图容器组件有自己的属性，如表 4.1 所示。

表 4.1 view 的属性

属性	类型	默认值	说明
hover-class	string	none	指定按下去的样式类。当 hover-class="none" 时，表示没有单击态效果
hover-stop-propagation	boolean	false	指定是否阻止本节点的祖先节点出现单击态
hover-start-time	number	50	指定按住后多久出现单击态，单位为 ms
hover-stay-time	number	400	指定手指松开后单击态的保留时间，单位为 ms

案例：在莫凡商城 index.wxml 文件页面里，输出"Hello World"文字、头像、昵称可以使用 view 布局，渲染出界面内容，如图 4.2 所示。

图 4.2 view 布局

具体代码如下。

```
<view class="container">
  <view class="userinfo">
    <image bindtap="bindViewTap" class="userinfo-avatar" src="https://wx.qlogo.cn/mmopen/vi_32/ Q0j4TwGTfTl249Dbdib9mqaKWK29vWLp2KlPHrMO0bwmOgLUE9T86XOG9kRx9PtBMRic4HFwqeHbUlK5IDWzvwPA/132" mode="cover"></image>
    <text class="userinfo-nickname">kevin</text>
  </view>
  <view class="usermotto">
    <text class="user-motto">Hello World</text>
  </view>
</view>
```

4.2.2 scroll-view 可滚动视图容器组件

scroll-view 是可滚动视图容器组件，允许视图容器中的内容进行横向滚动或者纵向滚动，类似于浏览器的水平滚动条和垂直滚动条，可以在有限的显示窗口中，通过滚动的方式显示更多的内容。scroll-view 的属性如表 4.2 所示。

表 4.2 scroll-view 的属性

属性	类型	默认值	说明
scroll-x	Boolean	false	允许横向滚动
scroll-y	Boolean	false	允许纵向滚动
upper-threshold	Number	50	指定距顶部/左边多远时，触发 scrolltoupper 事件，单位为像素
lower-threshold	Number	50	指定距底部/右边多远时，触发 scrolltolower 事件，单位为像素
scroll-top	Number		设置竖向滚动条位置
scroll-left	Number		设置横向滚动条位置
scroll-into-view	String		值应为某子元素 id，则滚动到该元素，元素顶部对齐滚动区域顶部
scroll-with-animation	Boolean	false	在设置滚动条位置时使用动画过渡
enable-back-to-top	Boolean	false	iOS 单击顶部状态栏；安卓双击标题栏时，滚动条返回顶部
bindscrolltoupper	eventHandle		滚动到顶部/左边，会触发 scrolltoupper 事件
bindscrolltolower	eventHandle		滚动到底部/右边，会触发 scrolltolower 事件
bindscroll	eventHandle		滚动时触发，event.detail = {scrollLeft, scrollTop, scrollHeight, scrollWidth, deltaX, deltaY}

1. 纵向滚动

要实现内容纵向滚动，需要给<scroll-view/>一个固定高度，在滑动的时候就会出现纵向的滚动条。

在莫凡商城 index.wxml 页面文件里增加纵向滚动条，如图 4.3 所示。

图 4.3 纵向滚动

具体代码如下。

```
<view class="container">
<scroll-view scroll-y="true" style="height: 200px;" bindscrolltoupper="upper" bindscrolltolower="lower">
  <view class="userinfo">
        <image bindtap="bindViewTap" class="userinfo-avatar" src="https://wx.qlogo.cn/mmopen/vi_32/Q0j4TwGTfTI249Dbdib9mqaKWK29vWLp2KlPHrMO0bwmOgLUE9T86XOG9kRx9PtBMRic4HFwqeHbUlK5IDWzvwPA/132" mode="cover"></image>
       <text class="userinfo-nickname">kevin</text>
   </view>
     </scroll-view>
<view class="usermotto">
    <text class="user-motto">Hello World</text>
</view>

</view>
```

2. 横向滚动

滴滴出行 App 在地图的上方显示可以打车的类别导航，有快车、单车、出租车、礼橙专车、公交、代驾、豪华车、安全须知、小桔租车、顺风车、关爱出行、车生活、金融服务，这些导航在一屏里无法完全显示出来，需要通过向左滑动和向右滑动显示完整的导航内容，单击相应的导航可以看到对应的内容，这时就可以采用 scroll-view 来实现这些导航的横向滚动，如图 4.4 所示。

下面模拟滴滴出行横向滚动效果，可以向左滑动和向右滑动，效果如图 4.5 所示。

 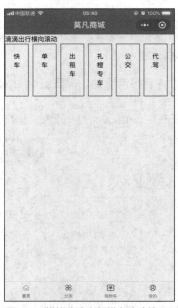

图 4.4　滴滴出行导航　　　　　　图 4.5　模拟滴滴出行横向滚动效果

在 WXML 文件里使用 scroll-view 进行布局，设置 scroll-x="true"横向滚动，具体代码如下。

```
<view class="section">
  <view class="section__title">滴滴出行横向滚动</view>
    <scroll-view scroll-x="true" style="width: 100%;">
      <view style="display:flex;flex-direction:row">
        <view style="margin-right:10px;border:1px solid blue;padding:20px;">快车</view>
        <view style="margin-right:10px;border:1px solid blue;padding:20px;">单车</view>
        <view style="margin-right:10px;border:1px solid blue;padding:20px;">出租车</view>
        <view style="margin-right:10px;border:1px solid blue;padding:20px;">礼橙专车</view>
```

```
            <view style="margin-right:10px;border:1px solid blue;padding:20px;">公交</view>
            <view style="margin-right:10px;border:1px solid blue;padding:20px;">代驾</view>
            <view style="margin-right:10px;border:1px solid blue;padding:20px;">豪华车</view>
            <view style="margin-right:10px;border:1px solid blue;padding:20px;">安全须知</view>
            <view style="margin-right:10px;border:1px solid blue;padding:20px;">小桔租车</view>
            <view style="margin-right:10px;border:1px solid blue;padding:20px;">顺风车</view>
            <view style="margin-right:10px;border:1px solid blue;padding:20px;">关爱出行</view>
            <view style="margin-right:10px;border:1px solid blue;padding:20px;">车生活</view>
            <view style="margin-right:10px;border:1px solid blue;padding:20px;">金融服务</view>
        </view>
    </scroll-view>
</view>
```

注意：

（1）基础库 2.4.0 以下不支持嵌套 textarea（多行输入框组件）、map（地图组件）、canvas（画布）和 video（视频）组件；

（2）scroll-into-view 的优先级高于 scroll-top。

（3）在滚动 scroll-view 时会阻止页面回弹，所以在 scroll-view 中滚动，是无法触发下拉刷新 onPullDownRefresh 的；若要使用下拉刷新，需要使用页面的滚动，而不是 scroll-view。

4.2.3　swiper 滑块视图容器组件

swiper 滑块视图容器组件是经常会用到的组件，它可以实现海报轮播效果或者多种登录方式（账号密码登录、手机号快捷登录）之间的切换，可以用来在指定区域内切换不同内容的显示，它的属性如表 4.3 所示。

表 4.3　swiper 的属性

属性	类型	默认值	说明
indicator-dots	boolean	false	设置是否显示面板指示点
indicator-color	color	rgba（0,0,0,.3）	设置指示点颜色
indicator-active-color	color	#000000	设置当前选中的指示点颜色
autoplay	boolean	false	设置是否自动切换
current	number	0	设置当前所在页面的 index
interval	number	5000	设置自动切换时间间隔
duration	number	500	设置滑动动画时长
circular	boolean	false	设置是否采用衔接滑动
vertical	boolean	false	设置滑动方向是否为纵向
previous-margin	string	"0px"	设置前边距，可用于露出前一项的一小部分，接受 px 和 rpx 值
next-margin	string	"0px"	设置后边距，可用于露出后一项的一小部分，接受 px 和 rpx 值
display-multiple-items	number	1	设置同时显示的滑块数量
skip-hidden-item-layout	boolean	false	设置是否跳过未显示的滑块布局，设为 true 可优化复杂情况下的滑动性能，但会丢失隐藏状态下滑块的布局信息
easing-function	string	"default"	指定 swiper 切换缓动动画类型
bindchange	eventHandle		current 改变时会触发 change 事件，event.detail = {current: current}
bindtransition	eventHandle		swiper-item 的位置发生改变时会触发 transition 事件，event.detail = {dx: dx, dy: dy}
bindanimationfinish	eventHandle		动画结束时会触发 animationfinish 事件，event.detail 同上

在 swiper 滑块视图容器组件里，嵌套有 swiper-item 组件，它用来显示不同页签的内容，一个 swiper 滑块视图容器组件里可以有多个 swiper-item 组件，来显示多个区域的内容，以实现海报轮播效果和页签切换效果。

1．海报轮播效果

海报轮播效果常用来展示商品图片信息或者广告信息。要在有限的区域内展示更多的内容，只能通过轮播的方式动态显示这些内容。海报轮播是网站和 App 都会采用的一种布局方式，如图 4.6 和图 4.7 所示。

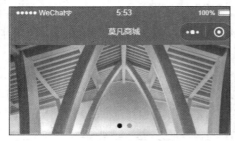 

图 4.6　海报 1　　　　　　　　　　　　图 4.7　海报 2

（1）在 WXML 文件里进行海报轮播区域的布局，采用 swiper 滑块视图容器组件进行布局，具体代码如下。

```
<view class="haibao">
    <swiper indicator-dots="{{indicatorDots}}" autoplay="{{autoplay}}" interval="{{interval}}" duration="{{duration}}">
        <block wx:for="{{imgUrls}}">
            <swiper-item>
                <image src="{{item}}" class="silde-image" style="width:100%"></image>
            </swiper-item>
        </block>
    </swiper>
</view>
```

（2）在 JS 文件里，提供海报轮播的图片、是否自动播放、轮播的时长等数据，通过数据绑定的方式渲染到页面上，具体代码如下。

```
Page({
  data: {
    indicatorDots: true,
    autoplay: true,
    interval: 5000,
    duration: 1000,
    imgUrls: [
      "../images/haibao/11.jpg",  "../images/haibao/22.jpg"
    ]
  }
})
```

设置 autoplay 等于 true 就可以自动进行海报轮播了，设置 indicatorDots 等于 true 则代表面板显示指示点，同时可以设置 interval 自动切换时长，duration 滑动动画时长。

使用 indicator-color 属性来设置指示点颜色，使用 indicator-active-color 属性来设置当前选中的指示点颜色，这样就可以根据自己的需求来设计更好的海报轮播效果。

2．页签切换效果

swiper 滑块视图容器组件除了可以实现海报轮播效果，还可以实现页签切换效果。它有一个 current

属性，表示当前所在页面的 index，根据 index 值来显示不同的页面，常用于多种方式的登录或者多种页签导航之间的切换，如图 4.8 和图 4.9 所示。

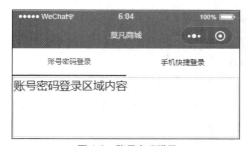 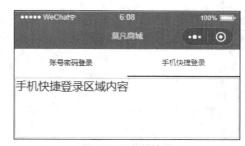

图 4.8　账号密码登录　　　　　　　　　　　图 4.9　手机快捷登录

（1）进入到 WXML 文件里，进行账号密码登录和手机快捷登录的界面布局设计，具体代码如下。

```
<view class="content">
    <view class="loginTitle">
        <view class="{{currentTab==0?'select':'default'}}" data-current="0" bindtap="switchNav">账号密码登录</view>
        <view class="{{currentTab==1?'select':'default'}}" data-current="1" bindtap="switchNav">手机快捷登录</view>
    </view>
    <view class="hr"></view>
    <swiper current="{{currentTab}}"style="height:120px">
        <swiper-item>
            <view style="margin:0 auto;border:1px solid #cccccc;width:99%;height:100px;">
                账号密码登录区域内容
            </view>
        </swiper-item>
        <swiper-item>
            <view style="margin:0 auto;border:1px solid #cccccc;width:99%;height:100px;">
                手机快捷登录区域内容
            </view>
        </swiper-item>
    </swiper>
</view>
```

（2）进入到 WXSS 文件里，给页面文件添加样式，具体代码如下。

```
.loginTitle{
    display: flex;
    flex-direction: row;
    width: 100%;
}
.select{
    font-size:12px;
    color: red;
    width: 50%;
    text-align: center;
    height: 45px;
    line-height: 45px;
    border-bottom:5rpx solid red;
}
.default{
    font-size:12px;
    margin: 0 auto;
    padding: 15px;
}
.hr{
    border: 1px solid #cccccc;
    opacity: 0.2;
}
```

（3）进入到 JS 文件里，提供当前面板的索引值，提供页签切换函数，具体代码如下。

```
Page({
  data: {
    currentTab: 0
  },
  switchNav: function (e) {
    var page = this;
    if (this.data.currentTab == e.target.dataset.current) {
      return false;
    } else {
      page.setData({ currentTab: e.target.dataset.current });
    }
  }
})
```

这样就可以实现在两种登录状态下的页签切换效果，页签切换时，页签的标题呈现为选中的状态，同时对应的内容也跟着进行切换。

4.2.4 movable-view 可移动视图容器组件

movable-view 是一个可移动视图容器组件，在页面中可以做拖曳滑动，在使用这个组件时，需要先定义可移动区域 movable-area，然后定义直接子节点 movable-view，否则不能移动。movable-area 要设置 width 和 height 属性，不设置时默认为 10 px。movable-view 也要设置 width 和 height 属性，不设置时默认为 10 px，movable-view 默认为绝对定位，top 和 left 属性为 0 px。movable-view 可移动视图容器的属性如表 4.4 所示。

表 4.4 movable-view 的属性

属性	类型	默认值	说明
direction	string	none	设置 movable-view 的移动方向，属性值有 all、vertical、horizontal、none
inertia	boolean	false	设置 movable-view 是否带有惯性
out-of-bounds	boolean	false	设置超过可移动区域后，movable-view 是否还可以移动
x	number / string		定义 x 轴方向的偏移，x 的值如果不在可移动范围内，会自动移动到可移动范围内；改变 x 的值会触发动画
y	number / string		定义 y 轴方向的偏移，y 的值如果不在可移动范围内，会自动移动到可移动范围内；改变 y 的值会触发动画
damping	number	20	阻尼系数，用于控制 x 或 y 改变时的动画和过界回弹的动画，值越大，移动的速度越快
friction	number	2	摩擦系数，用于控制惯性滑动的动画，值越大，摩擦力越大，滑动的速度越快停止；必须大于 0，否则会被设置成默认值
disabled	boolean	false	设置是否禁用
scale	boolean	false	设置是否支持双指缩放，默认缩放手势生效区域在 movable-view 内
scale-min	number	0.5	设置定义缩放倍数最小值
scale-max	number	10	设置定义缩放倍数最大值
scale-value	number	1	设置定义缩放倍数，取值范围为 0.5~10
animation	boolean	true	设置是否使用动画
bindchange	eventhandle		拖动过程中触发的事件，event.detail = {x: x, y: y, source: source}，其中 source 表示产生移动的原因，值可为 touch（拖动）、touch-out-of-bounds（超出移动范围）、out-of-bounds（超出移动范围后的回弹）、friction（惯性）和空字符串（setData）

续表

属性	类型	默认值	说明
bindscale	eventhandle		缩放过程中触发的事件，event.detail = {scale: scale}
htouchmove	eventhandle		初次手指触摸后移动为横向时触发，如果 catch 此事件，则意味着 touchmove 事件也被 catch
vtouchmove	eventhandle		初次手指触摸后移动为纵向时触发，如果 catch 此事件，则意味着 touchmove 事件也被 catch

movable-view 提供了 4 个事件：bindchange、bindscale、htouchmove 和 vtouchmove。

下面使用 movable-view 可移动视图容器组件来进行滑动，矩形区域代表可以移动的区域，其中的方块代表可以移动的组件，如图 4.10 所示。

图 4.10　可移动视图容器

（1）在 WXML 文件里使用 movable-area 和 movable-view 视图容器组件进行布局，具体代码如下。

```
<view class="section">
    <movable-area style="height: 200px; width:100%; background: yellow;">
        <movable-view style="height: 50px; width: 50px; background: red;" x="{{x}}" y="{{y}}" direction="all" bindchange="change" bindscale='scale' htouchmove="htouchmove" vtouchmove="vtouchmove">
        </movable-view>
    </movable-area>
</view>
```

（2）在 JS 文件里，提供拖动函数、缩放函数、初次手指触摸后移动为横向时触发函数、初次手指触摸后移动为纵向时触发函数，通过数据绑定的方式渲染到页面上，具体代码如下。

```
Page({
  data: {
    x: 0,
    y: 0
  },
  change: function (e) {
    console.log("拖动过程中触发的事件");
    console.log(e.detail)
  },
  scale: function (e) {
    console.log("缩放过程中触发的事件");
    console.log(e.detail)
  },
  htouchmove: function (e) {
    console.log("初次手指触摸后移动为横向时触发事件");
    console.log(e.detail)
  },
  vtouchmove: function (e) {
    console.log("初次手指触摸后移动为纵向时触发事件");
    console.log(e.detail)
  }
})
```

（3）当拖动可以移动的组件时，拖动过程中触发的事件就会触发，如拖动过程中，我们打印出拖动位置的日志，如图4.11所示。

图4.11 拖动时打印日志

4.2.5 cover-view 覆盖原生组件的视图容器组件

cover-view 和 cover-image 是覆盖原生组件的视图容器组件。例如，在使用地图组件时，本身地图组件功能有局限，我们想放置一些特殊的内容或图片，就需要使用到覆盖地图组件的视图容器。

（1）cover-view 可以在原生组件之上覆盖文本视图，可覆盖的原生组件包括 map、video、canvas、camera，只支持嵌套 cover-view、cover-image。

（2）cover-image 可以在原生组件之上覆盖图片视图，可覆盖的原生组件与 cover-view 相同，支持嵌套在 cover-view 中。

下面使用 cover-view、cover-image 覆盖原生组件的视图组件，在 video 视频播放组件上放置播放、暂停两个图片，同时放置一个时间内容显示区域，如图4.12 和图4.13 所示。

图4.12 视频播放

图4.13 覆盖视频播放组件

（1）在 WXML 文件里使用 cover-view、cover-image 覆盖原生组件的视图组件进行布局，具体代码如下。

```
    <video id="myVideo" src="http://wxsnsdy.tc.qq.com/105/20210/snsdyvideodownload?filekey=30280201010421301f
0201690402534804102ca905ce620b1241b726bc41dcff44e00204012882540400&bizid=1023&hy=SH&fileparam=302
c02010104253023020413fffd93020457e3c4ff02024ef202031e8d7f02030f42400204045a320a0201000400"
controls="{{false}}" event-model="bubble" style="width:100%">
      <cover-view class="controls">
        <cover-view class="play" bindtap="play">
          <cover-image class="img" src="../images/icon/play.jpg" />
        </cover-view>
        <cover-view class="pause" bindtap="pause">
          <cover-image class="img" src="../images/icon/pause.jpg" />
        </cover-view>
```

```
    <cover-view class="time">00:00</cover-view>
  </cover-view>
</video>
```

(2)在 WXSS 样式文件里添加样式,具体代码如下。

```
.controls {
  position: relative;
  top: 50%;
  height: 50px;
  margin-top: -25px;
  display: flex;
}
.play, .pause, .time {
  flex: 1;
  height: 100%;
}
.time {
  text-align: center;
  background-color: rgba(0, 0, 0, .5);
  color: white;
  line-height: 50px;
}
.img {
  width: 40px;
  height: 40px;
  margin: 5px auto;
}
```

(3)在 JS 文件里,提供视频播放、暂停函数,初始化视频播放组件,具体代码如下。

```
Page({
  onReady() {
    this.videoCtx = wx.createVideoContext('myVideo')
  },
  play() {
    this.videoCtx.play()
  },
  pause() {
    this.videoCtx.pause()
  }
})
```

4.2.6　项目实战:任务 9——实现搜索区域布局与海报轮播功能

1. 任务目标

通过实现搜索区域布局与海报轮播功能,学会利用 view 视图容器组件、image 图片组件来完成搜索区域布局,可以实现搜索区域布局水平居中和垂直居中;学会利用 swiper 滑块视图容器组件实现海报轮播功能。

莫凡商城首页顶部放置搜索区域和海报轮播区域,海报轮播区域可以动态地显示不同海报轮播内容,如图 4.14 和图 4.15 所示。

2. 任务实施

下面来实现首页搜索区域布局设计和海报轮播效果设计。

(1)在 index.wxml 文件里进行搜索区域布局设计,具体代码如下。

```
<view class="content">
  <view class="search">
    <view class="searchInput" bindtap="searchInput">
```

```
            <image src="/pages/images/tubiao/fangdajing-1.jpg" style="width:15px;height:19px;"></image>
            <text class="searchContent">搜索莫凡商品</text>
        </view>
    </view>
</view>
```

图 4.14　海报轮播效果 1

图 4.15　海报轮播效果 2

（2）在 index.wxss 文件里进行搜索区域布局样式渲染，具体代码如下。

```
.content{
    width: 100%;
    font-family: "Microsoft YaHei";
}
.search{
    width: 100%;
    background-color: #009966;
    height: 50px;
    line-height: 50px;
}
.searchInput{
    width: 95%;
    background-color: #ffffff;
    height: 30px;
    line-height: 30px;
    border-radius: 15px;
    display: flex;
    justify-content:center;
    align-items:center;
    margin: 0 auto;
}
.searchContent{
    font-size:12px;
    color: #777777;
}
```

（3）搜索区域界面布局的时候，使用了 view 组件、image 图片组件、text 组件，界面效果如图 4.16 所示。

图 4.16　搜索区域布局设计

（4）在 index.wxml 文件里进行海报轮播效果布局设计，具体代码如下。

```
<view class="content">
  <view class="search">
    <view class="searchInput" bindtap="searchInput">
      <image src="/pages/images/tubiao/fangdajing-1.jpg" style="width:15px;height:19px;"></image>
      <text class="searchContent">搜索莫凡商品</text>
    </view>
  </view>
  <view class="haibao">
    <swiper indicator-dots="{{indicatorDots}}" autoplay="{{autoplay}}" interval="{{interval}}" duration="{{duration}}">
      <block wx:for="{{imgUrls}}">
        <swiper-item>
          <image src="{{item}}" class="silde-image" mode="scaleToFill"></image>
        </swiper-item>
      </block>
    </swiper>
  </view>
</view>
```

（5）在 index.wxss 文件里进行海报轮播效果样式渲染，具体代码如下。

```
.content{
    width: 100%;
    font-family: "Microsoft YaHei";
}
.search{
    width: 100%;
    background-color: #009966;
    height: 50px;
    line-height: 50px;
}
.searchInput{
    width: 95%;
    background-color: #ffffff;
    height: 30px;
    line-height: 30px;
    border-radius: 15px;
    display: flex;
    justify-content:center;
    align-items:center;
    margin: 0 auto;
}
.searchContent{
    font-size:12px;
    color: #777777;
}
.haibao{
    text-align: center;
    width: 100%;
}
.silde-image{
    width: 100%;
}
```

（6）在 index.js 文件里进行海报轮播效果数据初始化，具体代码如下。

```
Page({
  data: {
    indicatorDots: true,
    autoplay: true,
    interval: 5000,
    duration: 1000,
    imgUrls: [
      "/pages/images/haibao/1.jpg",
      "/pages/images/haibao/2.jpg",
      "/pages/images/haibao/3.jpg"
    ]
  }
})
```

这样就完成了海报轮播效果界面布局、界面渲染、页面数据初始化及绑定操作，海报轮播图可以动态地轮播显示不同的内容，如图 4.17 所示。

图 4.17　海报轮播区域设计

在设计海报轮播区域的时候，我们用到了 view 视图容器组件、swiper 滑块视图容器组件，利用 swiper 滑块视图容器组件来实现海报轮播效果，用到了 swiper 的 indicatorDots（是否显示指示点）、autoplay（是否自动轮播）、interval（间隔时长）、duration（滑动时长）属性。在渲染列表的时候，我们用到了 wx: for 列表渲染。swiper 滑块视图容器组件还有更多的属性，我们可以自己去尝试使用。

4.3　基础内容组件

使用小程序的基础内容组件，我们能快速进行各种页面的布局设计。基础内容组件包括 icon 图标组件、text 文本组件、progress 进度条组件、progress 进度条组件、editor 富文本编辑器等。

慕课视频 image 图片组件及图片 API 应用

4.3.1　icon 图标组件

微信小程序提供了丰富的图标组件，应用于不同的场景，有成功、警告、提示、取消、下载等不同的含义，如图 4.18 所示。

icon 图标组件有 3 个属性：图标的类型 type、图标的大小 size 和图标的颜色 color，如表 4.5 所示。

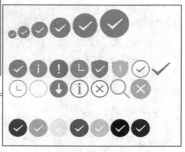

图 4.18　图标

表 4.5 icon 的属性

属性	类型	默认值	说明
type	string		icon 的类型，有效值：success、success_no_circle、info、warn、waiting、cancel、download、search、clear
size	number	23	icon 的大小，单位为像素
color	color		icon 的颜色，与 CSS 的 color 相同

下面使用 icon 组件绘制出如图 4.18 所示的图标。

（1）使用 icon 组件绘制不同尺寸的图标。

```
<view class="group">
    <icon type="success" size="20"/>
    <icon type="success" size="50"/>
    <icon type="success" size="60"/>
    <icon type="success" size="80"/>
    <icon type="success" size="100"/>
</view>
```

效果如图 4.19 所示。

（2）使用 icon 组件绘制不同类型的图标。

```
<view class="group">
    <icon type="success" size="45"/>
    <icon type="info" size="45"/>
    <icon type="warn" size="45"/>
    <icon type="waiting" size="45"/>
    <icon type="safe_success" size="45"/>
    <icon type="success_circle" size="45"/>
    <icon type="success_no_circle" size="45"/>
    <icon type="waiting_circle" size="45"/>
    <icon type="circle" size="45"/>
    <icon type="download" size="45"/>
    <icon type="info_circle" size="45"/>
    <icon type="cancel" size="45"/>
    <icon type="search" size="45"/>
    <icon type="clear" size="45"/>
</view>
```

效果如图 4.20 所示。

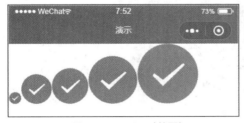

图 4.19 不同尺寸的图标

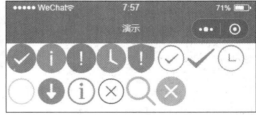

图 4.20 不同类型的图标

（3）使用 icon 组件绘制不同颜色的图标。

```
<view class="group">
    <icon type="success" size="45" color="red" />
    <icon type="success" size="45" color="orange" />
    <icon type="success" size="45" color="yellow" />
    <icon type="success" size="45" color="green" />
    <icon type="success" size="45" color="rgb(0, 255, 255)" />
    <icon type="success" size="45" color="blue" />
    <icon type="success" size="45" color="purple" />
</view>
```

效果如图 4.21 所示。

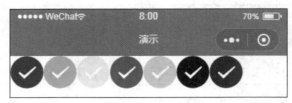

图 4.21　不同颜色的图标

这样就可以绘制出不同大小、不同类型、不同颜色的图标了，我们可以根据自己的需求，利用 icon 组件来设计小程序的图标。

4.3.2　text 文本组件

text 文本组件是用来放置文本信息的组件，它的属性如表 4.6 所示。

表 4.6　text 的属性

属性	类型	默认值	说明
selectable	boolean	false	文本是否可选
space	string	23	显示连续空格，ensp 表示中文字符空格一半大小，emsp 表示中文字符空格大小，nbsp 表示根据字体设置的空格大小
decode	boolean	false	是否解码

text 文本组件支持转义符"\"，如换行\n、空格\t，<text> 组件内只支持嵌套<text>组件，除了文本组件外的其他组件都无法长按选中。decode 属性可以解析的有不换行空格(   )、小于号(<)、大于号（>）、&符号（&）、引号（'）、半角空格（ ）、全角空格（ ），各个操作系统的空格标准并不一致。

使用转义符的示例代码如下。

```
<view class="btn-area">
  <view class="body-view">
    <text>我要学习\t 微信小程序</text>
    <text>我要成为\n 一名优秀工程师</text>
  </view>
  <view class="body-view">
    <text>我要学习\t 微信小程序</text>
  </view>
  <view class="body-view">
    <text>我要成为\n 一名优秀工程师</text>
  </view>
</view>
```

界面效果如图 4.22 所示。

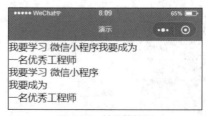

图 4.22　转义符效果

从图 4.21 中可以看出，\t 具有空格功能，\n 具有换行功能，同时也可以看出 text 文本组件是放置在一行里的，不同于 view 组件，每个 view 组件成一行。

4.3.3 progress 进度条组件

progress 进度条组件是一种用来提高用户体验度的组件，就像视频播放一样，可以通过进度条看到完整视频的长度、当前播放的进度，这样让用户能合理地安排自己的时间，提升用户体验，微信小程序也提供了 progress 进度条组件，它的属性如表 4.7 所示。

表 4.7 进度条属性

属性	类型	默认值	说明
percent	number	无	百分比，值为 0～100
show-info	boolean	false	在进度条右侧显示百分比
border-radius	number/string	0	圆角大小
font-size	number/string	16	右侧百分比字体大小
stroke-width	number	6	进度条线的宽度，单位为像素
color	string	#09BB07	进度条的颜色
activeColor	string	#09BB07	已选择的进度条的颜色
backgroundColor	string	#EBEBEB	未选择的进度条的颜色
active	boolean	false	进度条从左往右的动画
active-mode	string	backwards	backwards 表示动画从头播；forwards 表示动画从上次结束点接着播
bindactiveend	eventhandle		动画完成事件

可以尝试各种进度条的效果，示例代码如下。

```
<progress percent="20" show-info />
<progress percent="40" stroke-width="12" />
<progress percent="60" color="pink" />
<progress percent="80" active />
<progress percent="70" show-info stroke-width="20" border-radius="10" font-size="20" color="#CCCCCC" activeColor="#FF4040" backgroundColor="#6E8B3D" active active-mode="backwards" />
```

界面效果如图 4.23 所示。

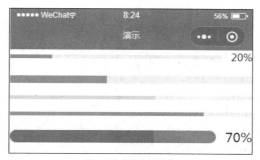

图 4.23 进度条效果

4.3.4 rich-text 富文本组件

通过 rich-text 富文本组件可以在 WXML 页面文件中显示一些富文本内容，如显示 HTML 的一些元素内容。rich-text 的属性如表 4.8 所示。

表 4.8 rich-text 的属性

属性	类型	默认值	说明
nodes	array/string	[]	节点列表/HTML
space	string		显示连续空格，ensp 为中文字符空格一半大小、emsp 为中文字符空格大小、nbsp 为根据字体设置的空格大小

rich-text 的 nodes 节点列表属性推荐使用 Array 类型。nodes 支持两种节点，通过 type 来区分，分别是元素节点和文本节点，默认为元素节点，即在 rich-text 富文本区域里显示的 HTML 节点。

1. 元素节点（type= node）

元素节点的属性如表 4.9 所示，受信任的 HTML 节点包括 a、abbr、address、article、aside、b、bdi、bdo、dir、big、blockquote、br、caption、center、cite、code、col、colgroup、dd、del、div、dl、dt、em、fieldset、font、footer、h1、h2、h3、h4、h5、h6、header、hr、i、img、ins、label、legend、li、mark、nav、ol、p、pre、q、rt、ruby、s、section、small、span、strong、sub、sup、table、tbody、td、tfoot、th、thead、tr、tt、u、ul。

表 4.9 元素节点的属性

属性	类型	默认值	说明
name	标签名	string	支持部分受信任的 HTML 节点
attrs	属性	Object	支持部分受信任的属性，遵循 Pascal 命名法
children	子节点列表	Array	结构和 nodes 一致

2. 文本节点（type= text）

文本节点的属性如表 4.10 所示。

表 4.10 文本节点的属性

属性	类型	默认值	说明
text	文本	string	支持 entities

示例代码如下。

```
<rich-text nodes="{{nodes}}" bindtap="tap"></rich-text>

Page({
  data: {
    nodes: [{
      name: 'div',
      attrs: {
        class: 'div_class',
        style: 'line-height: 60px; color: red;'
      },
      children: [{
        type: 'text',
        text: 'Hello World!'
      }]
    }]
  },
  tap() {
    console.log('tap')
  }
})
```

注意：

（1）nodes 不推荐使用 string 类型，如果使用 string 类型，组件会将 string 类型转换为 Array 类型，导致性能有所下降。

（2）rich-text 组件内屏蔽所有节点的事件。

（3）attrs 属性不支持 id，支持 class。

（4）name 属性对大小写不敏感。

（5）如果使用了不受信任的 HTML 节点，该节点及其所有子节点将会被移除。

（6）img 标签仅支持网络图片。

（7）如果在自定义组件中使用 rich-text 组件，那么仅自定义组件的 WXSS 样式对 rich-text 中的 class 生效。

4.3.5　editor 富文本编辑器及 API

editor 富文本编辑器，可以对图片、文字进行编辑，可以导出带标签的 html 和纯文本的 text 内容，富文本组件内部引入了一些基本的样式使得内容可以正确展示，开发时可以进行覆盖。需要注意的是，在其他组件或环境中使用富文本组件导出的 html 时，需要额外引入这段样式，并维护 <ql-container><ql-editor></ql-editor></ql-container> 的结构，editor 的属性如表 4.11 所示。

表 4.11　editor 的属性

属性	类型	默认值	说明
read-only	boolean	false	设置编辑器为只读
placeholder	string		提示信息
show-img-size	boolean	false	单击图片时显示图片大小控件
show-img-toolbar	boolean	false	单击图片时显示工具栏控件
show-img-resize	boolean	false	单击图片时显示修改尺寸控件
bindready	eventhandle		编辑器初始化完成时触发
bindfocus	eventhandle		编辑器聚焦时触发，event.detail = {html, text, delta}
bindblur	eventhandle		编辑器失去焦点时触发，detail = {html, text, delta}
bindinput	eventhandle		编辑器内容改变时触发，detail = {html, text, delta}
bindstatuschange	eventhandle		通过 Context 方法改变编辑器内样式时触发，返回选区已设置的样式

示例代码如下。

```
<view class="container">
  <view class="page-body">
    <editor id="editor" class="ql-container" placeholder="{{placeholder}}" bindready="onEditorReady" read-only="{{readOnly}}" bindinput="onContentChange" style="border:1px solid #cccccc;width:200px;" showImgSize showImgToolbar showImgResize>
    </editor>
    <view>
      <button bindtap="clickBtn">操作</button>
    </view>
  </view>
</view>

Page({
  data: {
    placeholder: '开始输入...',
    isReadOnly: false
  },
  onEditorReady:function() {//初始化编辑器
```

```
    var that = this;
    wx.createSelectorQuery().select('#editor').context(function (res) {
      that.editorCtx = res.context;
    }).exec()
  },
  onContentChange:function(e){//监控编辑器内容变化
    console.log(e.detail);
  },
  clickBtn:function(e) {//操作
    //清空编辑器内容
    this.editorCtx.clear();
    //插入文本内容
    this.editorCtx.insertText({
      text: "插入内容"
    });
    //插入图片
    this.editorCtx.insertImage({
      src: "https://api.mofun365.com:8888/images/banner/1555848473813.jpg"
    });
    //初始化编辑器内容
    this.editorCtx.setContents({
      html: "<h1>初始化编辑器内容<h1>"
    });
    //获取编辑器内容
    this.editorCtx.getContents({
      success:function(res) {
        console.log(res);
      }
    });
    //修改样式
    this.editorCtx.format("align", "center");
    //清除当前选区的样式
    this.editorCtx.removeFormat({
      success: function (res) {
        console.log("--------------------清除当前选区的样式------------------");
      }
    });
    //插入分割线
    this.editorCtx.insertDivider({
      success: function (res) {
        console.log("--------------------插入分割线------------------");
      }
    });
    //恢复
    this.editorCtx.redo({
      success: function (res) {
        console.log("--------------------恢复------------------");
      }
    });
    //撤销
    this.editorCtx.undo({
      success: function (res) {
        console.log("--------------------撤销------------------");
      }
    });
  }
})
```

编辑器效果如图 4.24 所示。

图 4.24　编辑器效果

（1）在 editor 组件上定义 id 属性，然后在 onEditorReady 函数里初始化富文本编辑器，获取 EditorContext 编辑器上下文对象，操作 editor 组件需要先将 EditorContext 实例化，具体代码如下。

```
onEditorReady:function() {//初始化编辑器
  var that = this;
  wx.createSelectorQuery().select('#editor').context(function (res) {
    that.editorCtx = res.context;
  }).exec()
}
```

（2）用 EditorContext.clear() 清空编辑器内容。

```
this.editorCtx.clear();
```

（3）用 EditorContext.insertText（Object object）插入文本内容，是覆盖当前选区内容，重新设置一段文本内容。

```
this.editorCtx.insertText({
    text: "插入内容"
})
```

（4）用 EditorContext.insertImage（Object object）插入图片，提供图片地址，仅支持 http（或 https）和 base64 格式。

```
this.editorCtx.insertImage({
    src: "https://api.mofun365.com:8888/images/banner/1555848473813.jpg"
});
```

（5）用 EditorContext. getContents() 获取编辑器内容。

```
this.editorCtx.getContents({
    success:function(res) {
      console.log(res);
    }
});
```

（6）用 EditorContext.format（string name, string value）修改样式。

```
this.editorCtx.format("align", "center");
```

（7）用 EditorContext.removeFormat() 清除当前选区的样式。

```
this.editorCtx.removeFormat({
    success: function (res) {
      console.log("--------------------清除当前选区的样式--------------------");
    }
});
```

（8）用 EditorContext.insertDivider() 插入分割线。

```
this.editorCtx.insertDivider({
    success: function (res) {
      console.log("--------------------插入分割线--------------------");
    }
});
```

（9）用 EditorContext.redo() 恢复之前的操作。

```
this.editorCtx.redo({
    success: function (res) {
```

```
        console.log("-------------------恢复-------------------");
    }
});
```
(10)用 EditorContext.undo()撤销之前操作。
```
this.editorCtx.undo({
    success: function (res) {
        console.log("-------------------撤销-------------------");
    }
});
```

4.4 image 图片组件及图片 API 的应用

4.4.1 image 图片组件

image 图片组件的默认宽度为 300 px、高度为 225 px。在 image 组件中,二维码、小程序码图片不支持长按识别,仅在 wx.previewImage 中支持长按识别。它有两类展现模式:一类是缩放模式,在缩放模式里包括 4 种方式;另一类是裁剪模式,在裁剪模式里包括 9 种方式,具体属性如表 4.12 所示。

表 4.12 image 的属性

属性	类型	默认值	说明
src	string		图片资源地址
mode	string	'scaleToFill'	图片裁剪、缩放的模式
lazy-load	boolean	false	图片懒加载,在即将进入一定范围(上下三屏)时才开始加载
show-menu-by-longpress	boolean	false	开启长按图片显示识别小程序码菜单
binderror	HandleEvent		当错误发生时,发布到 AppService 的事件名,事件对象 event.detail = {errMsg: 'something wrong'}
bindload	HandleEvent		当图片载入完毕时,发布到 AppService 的事件名,事件对象 event.detail = {height: '图片高度 px', width: '图片宽度 px'}

可以通过 mode 属性来设置 4 种缩放模式,如表 4.13 所示。

表 4.13 4 种缩放模式

模式	说明
scaleToFill	不保持纵横比缩放图片,使图片的宽、高完全拉伸至填满 image 元素
aspectFit	保持纵横比缩放图片,使图片的长边能完全显示出来。也就是说,可以完整地将图片显示出来
aspectFill	保持纵横比缩放图片,只保证图片的短边能完全显示出来。也就是说,图片通常只在水平或垂直方向是完整的,在另一个方向将会发生截取
widthFix	宽度不变,高度自动变化,保持原图宽、高比不变

可以通过 mode 属性来设置 9 种裁剪模式,如表 4.14 所示。

表 4.14 9 种裁剪模式

模式	说明
top	不缩放图片,只显示图片的顶部区域
bottom	不缩放图片,只显示图片的底部区域
center	不缩放图片,只显示图片的中间区域
left	不缩放图片,只显示图片的左边区域

续表

模式	说明
right	不缩放图片,只显示图片的右边区域
top left	不缩放图片,只显示图片的左上边区域
top right	不缩放图片,只显示图片的右上边区域
bottom left	不缩放图片,只显示图片的左下边区域
bottom right	不缩放图片,只显示图片的右下边区域

示例代码如下所示。

```
<view class="page">
  <view class="page__hd">
    <text class="page__title">image</text>
    <text class="page__desc">图片</text>
  </view>
  <view class="page__bd">
    <view class="section section_gap" wx:for="{{array}}" wx:for-item="item">
      <view class="section__title">{{item.text}}</view>
      <view class="section__ctn">
        <image style="width: 200px; height: 200px; background-color: #eeeeee;" mode="{{item.mode}}" src="{{src}}"></image>
      </view>
    </view>
  </view>
</view>
```

```
Page({
  data: {
    array: [{
      mode: 'scaleToFill',
      text: 'scaleToFill:不保持纵横比缩放图片,使图片完全适应'
    }, {
      mode: 'aspectFit',
      text: 'aspectFit:保持纵横比缩放图片,使图片的长边能完全显示出来'
    }, {
      mode: 'aspectFill',
      text: 'aspectFill:保持纵横比缩放图片,只保证图片的短边能完全显示出来'
    }, {
      mode: 'top',
      text: 'top:不缩放图片,只显示图片的顶部区域'
    }, {
      mode: 'bottom',
      text: 'bottom:不缩放图片,只显示图片的底部区域'
    }, {
      mode: 'center',
      text: 'center:不缩放图片,只显示图片的中间区域'
    }, {
      mode: 'left',
      text: 'left:不缩放图片,只显示图片的左边区域'
    }, {
      mode: 'right',
      text: 'right:不缩放图片,只显示图片的右边区域'
    }, {
      mode: 'top left',
      text: 'top left:不缩放图片,只显示图片的左上边区域'
```

```
    }, {
      mode: 'top right',
      text: 'top right:不缩放图片，只显示图片的右上边区域'
    }, {
      mode: 'bottom left',
      text: 'bottom left:不缩放图片，只显示图片的左下边区域'
    }, {
      mode: 'bottom right',
      text: 'bottom right:不缩放图片，只显示图片的右下边区域'
    }],
    src: '../images/icon/cat.jpg'
  },
  imageError: function(e) {
    console.log('image3 发生 error 事件，携带值为', e.detail.errMsg)
  }
})
```

图片效果如图 4.25～图 4.37 所示。

图 4.25　原图

图 4.26　scaleToFill 缩放模式

图 4.27　aspectFit 缩放模式

图 4.28　aspectFill 缩放模式

图 4.29　top 裁剪模式

图 4.30　bottom 裁剪模式

图 4.31　center 裁剪模式

图 4.32　left 裁剪模式

图 4.33　right 裁剪模式

图 4.34　top left 裁剪模式

图 4.35　top right 裁剪模式

图 4.36　bottom left 裁剪模式

图 4.37　bottom right 裁剪模式

4.4.2　图片 API 的应用

应用程序编程接口（API）是一种接口函数，可以让微信小程序访问应用程序，实现图片处理、文件操作、微信支付、分享等功能。

1. wx.chooseImage 选择图片 API

小程序通过 wx.chooseImage 选择图片 API 可以从本地相册选择图片或使用相机拍照来选择图片，wx.chooseImage 的参数说明如表 4.15 所示。

表 4.15　wx. chooseImage 的参数说明

属性	类型	是否必填	说明
count	number	否	最多可以选择的图片张数，默认为 9
sizeType	stringArray	否	设置显示 original 原图或 compressed 压缩图，默认为二者都有
sourceType	stringArray	否	album 为从相册选图，camera 为使用相机，默认为二者都有
success	Function	否	成功则返回图片的本地文件路径列表 tempFilePaths
fail	Function	否	接口调用失败的回调函数
complete	Function	否	接口调用结束的回调函数（调用成功、失败都会执行）

示例代码如下。

```
Page({
  onLoad:function(){
    wx.chooseImage({
      count: 9,  // 默认 9
      sizeType: ['original', 'compressed'],  // 可以指定是原图还是压缩图，默认二者都有
      sourceType: ['album', 'camera'],  // 可以指定来源是相册还是相机，默认二者都有
      success: function (res) {
        // 返回选定照片的本地文件路径列表，tempFilePath 可以作为 img 标签的 src 属性显示图片
```

```
          var tempFilePaths = res.tempFilePaths
      }
    })
  }
})
```

2. wx.previewImage 预览图片 API

wx.previewImage 预览图片 API 可以用来预览多张图片，并设置默认显示的图片，参数说明如表 4.16 所示。

表 4.16　wx. previewImage 的参数说明

属性	类型	是否必填	说明
current	string	否	当前显示图片的链接，不填则默认为 urls 的第一张
urls	stringArray	是	需要预览的图片链接列表
success	Function	否	接口调用成功的回调函数
fail	Function	否	接口调用失败的回调函数
complete	Function	否	接口调用结束的回调函数（调用成功、失败都会执行）

示例代码如下。

```
Page({
  onLoad:function(){
    wx.previewImage({
      current: 'http://img02.tooopen.com/images/20150928/tooopen_sy_143912755726.jpg', //当前显示图片的 http 链接
      urls: [
        "http://img02.tooopen.com/images/20150928/tooopen_sy_143912755726.jpg",
        "http://img06.tooopen.com/images/20160818/tooopen_sy_175866434296.jpg",
        "http://img06.tooopen.com/images/20160818/tooopen_sy_175833047715.jpg"
      ]// 需要预览的图片 http 链接列表
    })
  }
})
```

界面效果如图 4.38 和图 4.39 所示。

图 4.38　预览一

图 4.39　预览二

3. wx.getImageInfo 获得图片信息 API

wx.getImageInfo 用来获得图片信息，网络图片需先配置 download 域名才能生效，包括图片的宽度、图片的高度及图片返回的图片路径，参数说明如表 4.17 所示。

表 4.17 wx. getImageInfo 的参数说明

属性	类型	是否必填	说明
src	string	是	图片的路径，可以是相对路径、临时文件路径、存储文件路径、网络图片路径
success	Function	否	接口调用成功的回调函数
fail	Function	否	接口调用失败的回调函数
complete	Function	否	接口调用结束的回调函数（调用成功、失败都会执行）

success 返回参数说明如表 4.18 所示。

表 4.18 success 返回参数说明

参数	类型	说明
width	number	图片宽度，单位为像素
height	number	图片高度，单位为像素
path	string	返回图片的本地路径
orientation	string	拍照时设备的方向
type	string	图片格式

示例代码如下。

```
Page({
  onLoad:function(){
    wx.getImageInfo({
      src: 'http://img02.tooopen.com/images/20150928/tooopen_sy_143912755726.jpg',
      success: function (res) {
        console.log("图片宽度="+res.width);
        console.log("图片高度="+res.height);
        console.log("图片返回路径="+res.path);
      }
    })
  }
})
```

打印信息如图 4.40 所示。

图 4.40 打印信息

4. wx.saveImageToPhotosAlbum 保存图片到相册 API

微信小程序支持将图片保存到系统相册里，但需要用户授权，wx.saveImageToPhotosAlbum（OBJECT）的参数说明如表 4.19 所示。

表 4.19 wx.saveImageToPhotosAlbum 的参数说明

属性	类型	是否必填	说明
filePath	stringArray	是	图片文件路径，可以是临时文件路径，也可以是永久文件路径
success	Function	否	接口调用成功的回调函数
fail	Function	否	接口调用失败的回调函数
complete	Function	否	接口调用结束的回调函数（调用成功、失败都会执行）

wx.saveImageToPhotosAlbum 调用成功后会返回调用结果。

示例代码如下。

```
Page({
  data: {
    imgUrl: ''
  },
  onLoad: function () {
    var page = this;
    wx.downloadFile({
      url: "https://ss2.bdstatic.com/70cFvnSh_Q1YnxGkpoWK1HF6hhy/it/u=49292017,220644 01&fm=28&gp=0.jpg",
      type: 'image',
      success: function (res) {
        console.log(res);
        var tempPath = res.tempFilePath;

        wx.saveImageToPhotosAlbum({
          filePath: tempPath,
          success:function(res){//调用成功后返回调用结果
            console.log(res);
          }
        })
      }
    })
  }
})
```

5. wx.compressImage 压缩图片 API

微信小程序支持将图片进行压缩，图片压缩质量可以根据自己的需求对 quality 的属性进行设置，压缩质量的范围为 0～100，数值越小，压缩质量越低，压缩率越高（仅对 jpg 图片有效），wx.compressImage（OBJECT）参数说明如表 4.20 所示。

表 4.20 wx.compressImage 的参数说明

属性	类型	是否必填/默认值	说明
src	string	是	图片的路径，可以是相对路径、临时文件路径、存储文件路径
quality	number	80	压缩质量，范围为 0～100，数值越小，质量越低，压缩率越高（仅对 jpg 有效）
success	function	否	接口调用成功的回调函数
fail	function	否	接口调用失败的回调函数
complete	function	否	接口调用结束的回调函数（调用成功、失败都会执行）

wx.compressImage 调用成功后会返回压缩成功的临时路径 tempFilePath，微信开发者工具暂时不支持此 wx.compressImage 调试，需要使用真机进行开发调试。

示例代码如下。

```
Page({
  onLoad: function () {
    wx.compressImage({
```

```
      src: "https://ss2.bdstatic.com/70cFvnSh_Q1YnxGkpoWK1HF6hhy/it/u=49292017,22064401&fm=28&gp=0.jpg",  // 图片路径
      quality: 80,  // 压缩质量
      complete: function (res) {
        console.log(res);
      }
    })
  }
})
```

6. wx.chooseMessageFile 从客户端会话选择文件 API

微信小程序可以从客户端会话来选择文件,可以选择视频、图片及其他文件,wx.chooseMessageFile（OBJECT）参数说明如表 4.21 所示。

表 4.21　wx.chooseMessageFile 的参数说明

属性	类型	是否必填/默认值	说明
count	Number	是	最多可以选择的文件个数,范围为 0~100
type	String	'all'	所选文件的类型:all 为从所有文件选择,video 为只能选择视频文件,image 为只能选择图片文件,file 为可以选择除图片和视频之外的其他文件
extension	Array.<string>		根据文件拓展名过滤,仅 type==file 时有效。每一项都不能是空字符串。默认不过滤
fail	Function	否	接口调用失败的回调函数
complete	Function	否	接口调用结束的回调函数（调用成功、失败都会执行）
success	Function	否	接口调用成功的回调函数

wx.chooseMessageFile 从客户端会话选择文件调用成功后,会返回选择的文件的本地临时文件对象数组,对象数组包括 path（本地临时文件路径）、size（本地临时文件大小,单位为字节）、name（选择的文件名称）、type（选择的文件类型）、time（选择的文件的会话发送时间）。

示例代码如下。

```
Page({
  onLoad: function () {
    wx.chooseMessageFile({
      count: 10,
      type: 'image',
      success(res) {
        console.log(res);
      }
    })
  }
})
```

返回值打印信息如图 4.41 所示。

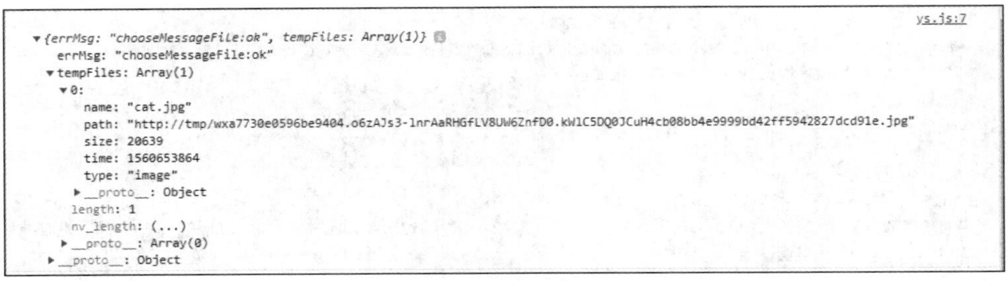

图 4.41　返回值打印信息

4.4.3 项目实战：任务 10（1）——实现图书列表显示功能静态布局

1. 任务目标

通过实现图书列表的静态布局，综合应用视图容器组件、基础内容组件、图片组件等组件，同时要学会进行页面布局及样式渲染。

图书列表包括 3 个区域：热门技术区域、秒杀时刻区域和畅销书籍区域，这 3 个区域布局方式一样，都是最多显示 3 本书，包括书籍图片、书籍名称、书籍价格，同时提供查看更多书籍入口，如图 4.42 和图 4.43 所示。

图 4.42 图书列表显示 1

图 4.43 图书列表显示 2

2. 任务实施

下面来实现图书列表显示功能静态布局。

（1）在 index.wxml 文件里进行图书列表布局，具体代码如下。

```
<view class="content">
<!--搜索区域-->
<view class="search">
    <view class="searchInput" bindtap="searchInput">
        <image src="/pages/images/tubiao/fangdajing-1.jpg" style="width:15px;height:19px;"></image>
        <text class="searchContent">搜索莫凡商品</text>
    </view>
</view>

<!--海报轮播区域-->
<view class="haibao">
    <swiper indicator-dots="{{indicatorDots}}" autoplay="{{autoplay}}" interval="{{interval}}" duration="{{duration}}">
        <block wx:for="{{imgUrls}}">
            <swiper-item>
                <image src="{{item}}" class="silde-image" mode="scaleToFill"></image>
```

```
      </swiper-item>
    </block>
  </swiper>
</view>

<!--图书列表区域-->
<view class="hr"></view>
<view class="list">
  <view class="tips">
    <view class="title">热门技术</view>
    <view class="more" bindtap='more' id="0">查看更多 ></view>
  </view>

  <view class="line"></view>
  <view class="items">
    <block wx:for="{{hotList}}">
      <view class="item" id="{{item.id}}" bindtap='seeDetail'>
        <view class="pic">
          <image src="{{item.listPic}}" mode="scaleToFill"></image>
        </view>
        <view class="name">{{item.goodsName}}</view>
        <view class="price">¥ {{item.goodsPrice}}</view>
      </view>
    </block>
  </view>
</view>
<view class="hr"></view>

<view class="list">
  <view class="tips">
    <view class="title">秒杀时刻</view>
    <view class="more" bindtap='more' id="1">查看更多 ></view>
  </view>

  <view class="line"></view>
  <view class="items">
    <block wx:for="{{spikeList}}">
      <view class="item" id="{{item.id}}" bindtap='seeDetail'>
        <view class="pic">
          <image src="{{item.listPic}}" mode="scaleToFill"></image>
        </view>
        <view class="name">{{item.goodsName}}</view>
        <view class="price">¥ {{item.goodsPrice}}</view>
      </view>
    </block>
  </view>
</view>
<view class="hr"></view>

<view class="list">
  <view class="tips">
    <view class="title">畅销书籍</view>
    <view class="more" bindtap='more' id="2">查看更多 ></view>
  </view>
  <view class="line"></view>
  <view class="items">
    <block wx:for="{{bestSellerList}}">
      <view class="item" id="{{item.id}}" bindtap='seeDetail'>
        <view class="pic">
          <image src="{{item.listPic}}" mode="scaleToFill"></image>
        </view>
        <view class="name">{{item.goodsName}}</view>
        <view class="price">¥ {{item.goodsPrice}}</view>
      </view>
```

```
        </block>
      </view>
    </view>
    <view class="hr"></view>
</view>
```

（2）在 index.wxss 文件里进行图书列表样式渲染，具体代码如下。

```css
.content{
    width: 100%;
    font-family: "Microsoft YaHei";
}
.search{
    width: 100%;
    background-color: #009966;
    height: 50px;
    line-height: 50px;
}
.searchInput{
    width: 95%;
    background-color: #ffffff;
    height: 30px;
    line-height: 30px;
    border-radius: 15px;
    display: flex;
    justify-content:center;
    align-items:center;
    margin: 0 auto;
}
.searchContent{
    font-size:12px;
    color: #777777;
}
.haibao{
    text-align: center;
    width: 100%;
}
.silde-image{
    width: 100%;
}
.list{
    height: 250px;
}
.line{
    height: 1px;
    width: 100%;
    background-color: #009966;
    opacity: 0.2;
}
.hr{
    height: 10px;
    width: 100%;
    background-color: #cccccc;
    opacity: 0.2;
}
.tips{
    display: flex;
    flex-direction: row;
}
.title{
    padding: 10px;
    font-size: 15px;
    color: #009966;
    font-weight: bold;
}
```

```css
.more{
    position: absolute;
    right: 10px;
    margin-top: 12px;
    font-size: 12px;
    color: #999999;
}
.items{
    padding: 10px;
    display: flex;
    flex-wrap: wrap;
    justify-content: space-left;
}
.item{
    width: 30%;
    height: 190px;
    border: 1px solid #009966;
    border-radius: 5px;
    margin-top: 10px;
    margin-bottom: 10px;
    margin: 0 auto;
    text-align: center;
}
.pic{
    margin-top: 5px;
}
.pic image{
    width:100px;
    height:120px;
}
.name{
    font-size: 12px
}
.price{
    color:red;
    margin-top:2px;
    font-size: 16px;
}
```

（3）在 index.js 文件里初始化热门书籍 hotList 数据、秒杀时刻书籍 spikeList 数据、畅销书籍 bestSellerList 数据，具体代码如下。

```
Page({
    data: {
        indicatorDots: true,
        autoplay: true,
        interval: 5000,
        duration: 1000,
        imgUrls: [
            "/pages/images/haibao/1.jpg",
            "/pages/images/haibao/2.jpg",
            "/pages/images/haibao/3.jpg"
        ],
        hotList: [
            { "id": 1,  "listPic": "https://api.mofun365.com:8888/images/goods/1555850845474.jpg",  "goodsName": "微信小程序开发图解案例教程",  "goodsPrice": 62.8},
            { "id": 2,  "listPic": "https://api.mofun365.com:8888/images/goods/1555851154057.jpg",  "goodsName": "微信小程序开发全案精讲",  "goodsPrice": 41.88 },
            { "id": 3,  "listPic": "https://api.mofun365.com:8888/images/goods/1555851345937.jpg",  "goodsName": "第一行代码 Java",  "goodsPrice": 57.7 }
        ],
        spikeList: [
            { "id": 4,  "listPic":  "https://api.mofun365.com:8888/images/goods/1555851497575.jpg",  "goodsName": "Android 原理解析与开发指南",  "goodsPrice": 35.99 },
            { "id": 5,  "listPic": "https://api.mofun365.com:8888/images/goods/1555851661073.png",  "goodsName": "响应
```

```
式 Web 开发项目教程", "goodsPrice": 36.4},
        { "id": 6, "listPic": "https://api.mofun365.com:8888/images/goods/1555851817322.jpg", "goodsName": "第一
行代码 C 语言", "goodsPrice": 41.99 }
    ],
    bestSellerList: [
        { "id": 7, "listPic": "https://api.mofun365.com:8888/images/goods/1555851965264.jpg", "goodsName": "前端
HTML+CSS 修炼之道", "goodsPrice": 57.7 },
        { "id": 8, "listPic": "https://api.mofun365.com:8888/images/goods/1555850845474.jpg", "goodsName": "微信
小程序开发图解案例教程", "goodsPrice": 62.8 },
        { "id": 9, "listPic": "https://api.mofun365.com:8888/images/goods/1555851154057.jpg", "goodsName": "微信
小程序开发全案精讲", "goodsPrice": 41.8 }
    ]
  }
})
```

这样就实现了图书列表显示功能静态布局。在 index.js 文件里初始化了一些书籍数据，在后面学习网络请求的时候，就可以通过网络请求来动态地获取书籍数据。在进行热门、秒杀时刻、畅销书籍区域布局的时候可以发现，它们布局的方式完全一致，这样就可以先设计一个区域，把一个区域完全设计好之后，可以通过拷贝的方式直接复制，在这个基础上改就会减轻很多工作量。

4.5 导航组件和导航 API

微信小程序可以在页面中设置导航，使用 navigator 页面链接组件，也可以在 JS 文件里通过导航 API 进行页面跳转，同时可以设置导航条标题和显示动画效果。

4.5.1 navigator 页面链接组件

navigator 页面链接组件是用在 WXML 页面中跳转的导航，它有以下 3 种类型。
（1）保留当前页跳转，跳转后可以返回当前页，它与 wx.navigateTo 跳转效果一样。
（2）关闭当前页跳转，无法返回当前页，它与 wx.redirectTo 跳转效果一样。
（3）跳转到底部标签导航指定的页面，它与 wx.switchTab 跳转效果一样。

navigator 页面链接组件的这些跳转效果都是通过 open-type 属性来控制的，具体属性如表 4.22 所示。

表 4.22 navigato 页面链接属性

属性	类型	默认值	说明
target	string	self	在哪个目标上发生跳转，默认当前小程序，self 当前小程序，miniProgram 其他小程序
url	string		应用内的跳转链接
open-type	string	Navigate	navigate 对应 wx.navigateTo 或 wx.navigateToMiniProgram 的功能； redirect 对应 wx.redirectTo 的功能； switchTab 对应 wx.switchTab 的功能； reLaunch 对应 wx.reLaunch 的功能； navigateBack 对应 wx.navigateBack 的功能； exit 退出小程序，target="miniProgram"时生效
delta	number	1	当 open-type 为'navigateBack'时有效，表示回退的层数
app-id	string		当 target="miniProgram"时有效，表示要打开的小程序 AppID
path	string		当 target="miniProgram"时有效，表示打开的页面路径，如果为空则打开首页

续表

属性	类型	默认值	说明
extra-data	object		当 target="miniProgram"时有效，表示需要传递给目标小程序的数据，目标小程序可在 App.onLaunch()、App.onShow()中获取到这份数据
version	string	release	当 target="miniProgram"时有效，表示要打开的小程序版本，有 develop 开发版、trial 体验版、release 正式版，仅在当前小程序为开发版或体验版时此参数有效；如果当前小程序是正式版，则打开的小程序必定是正式版
hover-class	string	navigator-hover	指定单击时的样式类，当 hover-class="none"时，没有单击态效果
hover-stop-propagation	boolean	false	指定是否阻止本节点的祖先节点出现单击态
hover-start-time	number	50	指定按住后多久出现单击态，单位为毫秒
hover-stay-time	number	600	指定手指松开后单击态的保留时间，单位为毫秒
bindsuccess	string		当 target="miniProgram"时有效，表示跳转小程序成功
bindfail	string		当 target="miniProgram"时有效，表示跳转小程序失败
bindcomplete	string		当 target="miniProgram"时有效，表示跳转小程序完成

下面来演示一下 open-type 不同导航类型的跳转效果。

（1）新建一个 navigator 项目，进入到 app.json 文件里，在 pages 属性里设置页面路径 "pages/index/index" "pages/navigator/navigator" "pages/redirect/redirect"，具体代码如下。

```
{
  "pages":[
    "pages/index/index",
    "pages/navigator/navigator",
    "pages/redirect/redirect"
  ],
  "window":{
    "backgroundTextStyle":"light",
    "navigationBarBackgroundColor": "#fff",
    "navigationBarTitleText": "导航",
    "navigationBarTextStyle":"black"
  },
}
```

（2）进入到 pages/index/index.wxml 文件里，设计导航的3种跳转方式：保留当前页跳转、关闭当前页跳转、跳转到 tabBar 页面，具体代码如下。

```
<view class="btn-area">
    <navigator url="../navigator/navigator?title=navigator" open-type="navigate" hover-class="navigator-hover">wx.navigateTo 保留当前页跳转</navigator>
    <navigator url="../redirect/redirect?title=redirect" open-type="redirect" hover-class="other-navigator-hover">wx.redirectTo 关闭当前页跳转</navigator>
    <navigator url="../redirect/redirect" open-type="switchTab" hover-class="other-navigator-hover">wx.switchTab 跳转到 tabBar 页面</navigator>
</view>
```

（3）进入到 pages/navigator/navigator.wxml 文件里，进行界面布局，具体代码如下。

```
<view>保留当前页进行跳转，单击左上角可以返回到当前页</view>
```

（4）进入到 pages/redirect/redirect.wxml 文件里，进行界面布局，具体代码如下。

```
<view>关闭当前页进行跳转，跳转后无法返回到当前页 </view>
```

（5）wx.navigateTo 保留当前页跳转和 wx.redirectTo 关闭当前页跳转都可以正常跳转，但是 wx.switchTab 跳转到 tabBar 页面无法完成跳转，它需要在 app.json 文件的 tabBar 属性里设置底部标签导航，具体代码如下。

```
{
  "pages":[
```

```
    "pages/index/index",
    "pages/navigator/navigator",
    "pages/redirect/redirect"
  ],
  "window":{
    "backgroundTextStyle":"light",
    "navigationBarBackgroundColor": "#fff",
    "navigationBarTitleText": "导航",
    "navigationBarTextStyle":"black"
  },
  "tabBar": {
    "selectedColor": "red",
    "list": [{
      "pagePath": "pages/index/index",
      "text": "首页",
      "iconPath": "iconPath",
      "selectedIconPath": "selectedIconPath"
    }, {
      "pagePath": "pages/redirect/redirect",
      "text": "当前页打开导航",
      "iconPath": "iconPath",
      "selectedIconPath": "selectedIconPath"
    }]
  }
}
```

（6）wx.switchTab 跳转到 tabBar 页面可以跳转到指定的底部标签导航页面里，但是可以发现用 wx.navigateTo 保留当前页跳转和 wx.redirectTo 关闭当前页跳转这两种方式无法跳转，因为在 app.json 文件中配置的 tabBar 属性里设置了底部标签导航。

（7）navigator 页面链接组件设置的跳转路径，如果带参数，如 url="../navigator/navigator?title=navigator"，则 title 的值可以在跳转页面的.js 文件的 onLoad 函数里获取，具体代码如下。

```
Page({
  data:{},
  onLoad:function(options){
    console.log("title="+options);
  }
})
```

4.5.2 wx.navigateTo 保留当前页跳转 API

wx.navigateTo 保留当前页面，跳转到应用内的某个页面，使用 wx.navigateBack 可以返回到原页面，小程序中页面栈最多为 10 层。具体属性如表 4.23 所示。

表 4.23 wx.navigateTo 的属性

属性	类型	是否必填	说明
url	string	是	需要跳转的应用内非 tabBar 的页面的路径，路径后可以带参数。参数与路径之间使用"?"分隔，参数键与参数值用"="相连，不同参数用"&"分隔，如'path?key=value&key2=value2'
success	Function	否	接口调用成功的回调函数
fail	Function	否	接口调用失败的回调函数
complete	Function	否	接口调用结束的回调函数（调用成功、失败都会执行）

（1）进入到 pages/index/index.wxml 文件里，添加一个"跳转"按钮，保留当前页进行跳转，具体代码如下。

```
<view class="btn-area">
  <navigator url="../navigator/navigator?title=navigator11" open-type="navigate" hover-class="navigator-hover">
```

```
wx.navigateTo 保留当前页跳转</navigator>
        <navigator  url="../redirect/redirect?title=redirect"  open-type="redirect"  hover-class="other-navigator-hover">
wx.redirectTo 关闭当前页跳转</navigator>
        <navigator url="../redirect/redirect" open-type="switchTab" hover-class="other-navigator-hover">wx.switchTab 跳
转到 tabBar 页面</navigator>
        <button type="primary" bindtap="navigateBtn">保留当前页跳转</button>
    </view>
```

（2）进入到 pages/index/index.js 文件里，添加一个 navigateBtn 事件函数，保留当前页并跳转到 pages/navigator/navigator.wxml 页面文件里，具体代码如下。

```
Page({
    navigateBtn:function(){
        wx.navigateTo({
            url: '../navigator/navigator',
            success: function(res){
                console.log(res);
            },
            fail: function() {
                // fail
            },
            complete: function() {
                // complete
            }
        })
    }
})
```

4.5.3 wx.redirectTo 关闭当前页跳转 API

wx.redirectTo 关闭当前页面，跳转到应用内的某个页面，但是不允许跳转到 tabbar 页面。具体属性如表 4.24 所示。

表 4.24 wx.redirectTo 的属性

属性	类型	是否必填	说明
url	string	是	需要跳转的应用内非 tabBar 的页面的路径，路径后可以带参数。参数与路径之间使用"？"分隔，参数键与参数值用"＝"相连，不同参数用"&"分隔，如'path?key=value&key2=value2'
success	Function	否	接口调用成功的回调函数
fail	Function	否	接口调用失败的回调函数
complete	Function	否	接口调用结束的回调函数（调用成功、失败都会执行）

（1）进入到 pages/index/index.wxml 文件里，添加一个"跳转"按钮，关闭当前页进行跳转，具体代码如下。

```
<view class="btn-area">
        <navigator url="../navigator/navigator?title=navigator11" open-type="navigate" hover-class="navigator-hover">
wx.navigateTo 保留当前页跳转</navigator>
        <navigator url="../redirect/redirect?title=redirect" open-type="redirect" hover-class="other-navigator-hover"> wx.redirectTo 关
闭当前页跳转</navigator>
        <navigator url="../redirect/redirect" open-type="switchTab" hover-class="other-navigator-hover">wx.switchTab 跳
转到 tabBar 页面</navigator>
        <button type="primary" bindtap="navigateBtn">保留当前页跳转</button>
        <button type="primary" bindtap="redirectBtn">关闭当前页跳转</button>
    </view>
```

（2）进入到 pages/index/index.js 文件里，添加一个 redirectBtn 事件函数，保留当前页并跳转到 pages/navigator/navigator.wxml 页面里，具体代码如下。

```
Page({
    navigateBtn:function(){
```

```
        wx.navigateTo({
          url: '../navigator/navigator',
          success: function(res){
            console.log(res);
          },
          fail: function() {
            // fail
          },
          complete: function() {
            // complete
          }
        })
      },
      redirectBtn:function(){
        wx.redirectTo({
          url: '../navigator/navigator',
          success: function(res){
            console.log(res);
          },
          fail: function() {
            // fail
          },
          complete: function() {
            // complete
          }
        })
      }
})
```

4.5.4 wx.switchTab 跳转到 tabBar 页面 API

跳转到 tabBar 页面，并关闭其他所有非 tabBar 页面，具体属性如表 4.25 所示。

表 4.25 wx.switchTab 的属性

属性	类型	是否必填	说明
url	string	是	需要跳转的 tabBar 页面的路径（需在 App.json 的 tabBar 字段定义的页面），路径后不能带参数
success	Function	否	接口调用成功的回调函数
fail	Function	否	接口调用失败的回调函数
complete	Function	否	接口调用结束的回调函数（调用成功、失败都会执行）

（1）进入到 pages/index/index.wxml 文件里，添加一个"跳转"按钮，跳转到 tabBar 页面，具体代码如下。

```
<view class="btn-area">
    <navigator url="../navigator/navigator?title=navigator11" open-type="navigate" hover-class="navigator-hover">wx.navigateTo 保留当前页跳转</navigator>
    <navigator url="../redirect/redirect?title=redirect" open-type="redirect" hover-class="other-navigator-hover">wx.redirectTo 关闭当前页跳转</navigator>
    <navigator url="../redirect/redirect" open-type="switchTab" hover-class="other-navigator-hover">wx.switchTab 跳转到 tabBar 页面</navigator>
    <button type="primary" bindtap="navigateBtn">保留当前页跳转</button>
    <button type="primary" bindtap="redirectBtn">关闭当前页跳转</button>
    <button type="primary" bindtap="switchBtn">跳转到 tabBar 页面</button>
</view>
```

（2）进入到 pages/index/index.js 文件里，添加一个 navigateBtn 事件函数，保留当前页并跳转到 pages/redirect/redirect.wxml 页面里，具体代码如下。

```
Page({
    navigateBtn:function(){
        wx.navigateTo({
            url: '../navigator/navigator',
            success: function(res){
                console.log(res);
            },
            fail: function() {
                // fail
            },
            complete: function() {
                // complete
            }
        })
    },
    redirectBtn:function(){
        wx.redirectTo({
            url: '../navigator/navigator',
            success: function(res){
                console.log(res);
            },
            fail: function() {
                // fail
            },
            complete: function() {
                // complete
            }
        })
    },
    switchBtn:function(){
        wx.switchTab({
            url: '../redirect/redirect',
            success: function(res){
                // success
            },
            fail: function() {
                // fail
            },
            complete: function() {
                // complete
            }
        })
    }
})
```

wx.navigateTo 和 wx.redirectTo 不允许跳转到 tabbar 页面，只能用 wx.switchTab 跳转到 tabBar 页面。

4.5.5 wx.navigateBack 返回上一页 API

wx.navigateBack 关闭当前页面，返回上一页面或多级页面，可通过 getCurrentPages()) 获取当前的页面栈，决定需要返回几层。具体属性如表 4.26 所示。

表 4.26 wx.navigateBack 的属性

属性	类型	是否必填	说明
delta	number	1	返回的页面数，如果 delta 大于现有页面数，则返回到首页

（1）进入到 pages/navigator/navigator.wxml 文件里，添加一个"返回"按钮，单击"返回"按

钮，可以返回到上一级页面，具体代码如下。

```
<view>保留当前页进行跳转，单击左上角可以返回到当前页</view>
<button type="primary" bindtap="backBtn">返回上一页</button>
```

（2）进入到 pages/navigator/navigator.js 文件里，添加 backBtn 事件返回函数，具体代码如下。

```
Page({
  data:{},
  onLoad:function(options){
    console.log("title="+options);
  },
  backBtn:function(){
    wx.navigateBack({
      delta: 1
    })
  }
})
```

（3）在 pages/index/index.wxml 文件里，单击"保留当前页跳转"按钮，可以进行页面跳转，在跳转的页面里单击"返回上一页"按钮，可以返回到上一级页面，如图 4.44 和图 4.45 所示。

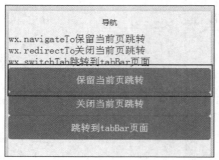
图 4.44　index.wxml 页面

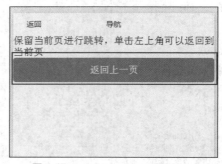
图 4.45　navigator.wxml 页面

4.5.6　wx.reLaunch 关闭所有页面，打开某个页面 API

wx. reLaunch 是关闭所有页面，然后打开应用内的某个页面的跳转方式。具体属性如表 4.27 所示。

表 4.27　wx. reLaunch 的属性

属性	类型	是否必填	说明
url	string	是	需要跳转的应用内的页面路径，路径后可以带参数。参数与路径之间使用"?"分隔，参数键与参数值用"="相连，不同参数用"&"分隔，如'path?key=value&key2=value2
success	Function	否	接口调用成功的回调函数
fail	Function	否	接口调用失败的回调函数
complete	Function	否	接口调用结束的回调函数（调用成功、失败都会执行）

示例代码如下。

```
wx.reLaunch({
  url: 'test?id=1'
})
```

4.5.7　导航条 API

导航条 API 有 4 种：wx.showNavigationBarLoading（Object object）在当前页面显示导航条加载动画、wx.hideNavigationBarLoading（Object object）在当前页面隐藏导航条加载动画、

wx.setNavigationBarTitle（Object object）动态设置当前页面的标题、wx.setNavigationBarColor（Object object）设置页面导航条颜色。

示例代码如下。

```
Page({
  onLoad: function (options) {
    //在当前页面显示导航条加载动画
    wx.showNavigationBarLoading({
      success:function(){
        console.log("在当前页面显示导航条加载动画");
      }
    });

    //在当前页面隐藏导航条加载动画
    wx.hideNavigationBarLoading({
      success: function () {
        console.log("在当前页面隐藏导航条加载动画");
      }
    });

    //动态设置当前页面的标题
    wx.setNavigationBarTitle({
      title: '新页面',
      success:function(){
        console.log("动态设置当前页面的标题");
      }
    });

    //设置页面导航条颜色
    wx.setNavigationBarColor({
      frontColor: '#ffffff',   //前景颜色值，包括按钮、标题、状态栏的颜色，仅支持 #ffffff 和 #000000
      backgroundColor: '#ff0000',//背景颜色值，有效值为十六进制颜色
      animation: { //动画效果
        duration: 400,   //动画变化时间
        timingFunc: 'easeIn' ////动画变化方式
      },
      success: function () {
        console.log("设置页面导航条颜色");
      }
    });
  }
})
```

wx.showNavigationBarLoading（Object object）在当前页面显示导航条加载动画、wx.hideNavigationBarLoading（Object object）在当前页面隐藏导航条加载动画、wx.setNavigationBarTitle（Object object）动态设置当前页面的标题这 3 个 API 使用起来比较简单，按照示例代码使用即可。

wx.setNavigationBarColor（Object object）设置页面导航条颜色稍微复杂一点，具体属性如表 4.28 所示。

表 4.28 wx. setNavigationBarColor 的属性

属性	类型	是否必填	说明
frontColor	String	是	前景颜色值，包括按钮、标题、状态栏的颜色，仅支持#ffffff 和#000000
backgroundColor	String	是	背景颜色值，有效值为十六进制颜色
Animation	Object	是	动画效果
Success	Function	否	接口调用成功的回调函数
Fail	Function	否	接口调用失败的回调函数
Complete	Function	否	接口调用结束的回调函数（调用成功、失败都会执行）

animation 动画效果对象包含两个属性：duration 动画变化时间，单位为毫秒；timingFunc 动画变化方式，提供了 linear 动画从头到尾的速度是相同的、easeIn 动画以低速开始、easeOut 动画以低速结束、easeInOut 动画以低速开始和结束 4 种变化方式，示例代码如下。

```
wx.setNavigationBarColor({
    frontColor: '#ffffff',   //前景颜色值，包括按钮、标题、状态栏的颜色，仅支持 #ffffff 和 #000000
    backgroundColor: '#ff0000', //背景颜色值，有效值为十六进制颜色
    animation: { //动画效果
        duration: 400,    //动画变化时间
        timingFunc: 'easeIn' //动画变化方式
    },
    success: function () {
        console.log("设置页面导航条颜色");
    }
});
```

4.5.8 Tab Bar 标签导航 API

为灵活处理 Tab Bar 标签导航，微信小程序提供了 8 个有关标签导航的 API。
（1）wx.showTabBarRedDot（Object object）：显示 tabBar 某一项右上角的红点。
（2）wx.hideTabBarRedDot（Object object）：隐藏 tabBar 某一项右上角的红点。
（3）wx.showTabBar（Object object）：显示 tabBar 标签导航。
（4）wx.hideTabBar（Object object）：隐藏 tabBar 标签导航。
（5）wx.setTabBarStyle（Object object）：动态设置 tabBar 的整体样式。
（6）wx.setTabBarItem（Object object）：动态设置 tabBar 某一项的内容。
（7）wx.setTabBarBadge（Object object）：为 tabBar 某一项的右上角添加文本。
（8）wx.removeTabBarBadge（Object object）：移除 tabBar 某一项右上角的文本。

1. 在 tabBar 某一项的右上角显示红点

wx.showTabBarRedDot（Object object）和 wx.hideTabBarRedDot（Object object）是比较常用的功能，可以在 tabBar 右上角显示或隐藏红点，当有新消息时就可以通过这种方式进行提醒，它有一个 index 属性，从左边算起，index 值从 0 开始算起，根据 index 值来设置哪一个 tabBar 右上角显示或隐藏红点。

示例代码如下。

```
onLoad:function(){
    wx.showTabBarRedDot({
        index:0
    });
    wx.hideTabBarRedDot({
        index:1
    });
}
```

效果图如图 4.46 所示。

图 4.46 tabBar 右上角显示或隐藏红点

2. 显示或隐藏 tabBar 标签导航

wx.showTabBar（Object object）和 wx.hideTabBar（Object object）可以动态地控制 tabBar

标签导航的显示或隐藏。

示例代码如下。

```
onLoad:function(){
  wx.showTabBar({
    animation:true //是否需要动画效果
  });
  wx. hideTabBar ({
    animation:true //是否需要动画效果
  });
}
```

3. 动态设置 tabBar 整体样式

wx.setTabBarStyle（Object object）可以动态地设置 tabBar 样式，参数 Object 对象的具体属性如表 4.29 所示。

表 4.29　wx.setTabBarStyle 的属性

属性	类型	是否必填	说明
color	string	是	tab 上文字的默认颜色，HexColor
selectedColor	string	是	tab 上的文字选中时的颜色，HexColor
backgroundColor	string	是	tab 的背景色，HexColor
borderStyle	string	是	tabBar 上边框的颜色，仅支持 black/white
success	Function	否	接口调用成功的回调函数
fail	Function	否	接口调用失败的回调函数
complete	Function	否	接口调用结束的回调函数（调用成功、失败都会执行）

示例代码如下。

```
wx.setTabBarStyle({
  color: '#FF0000',
  selectedColor: '#00FF00',
  backgroundColor: '#0000FF',
  borderStyle: 'white'
})
```

4. 动态设置 tabBar 某一项内容

wx.setTabBarItem（Object object）可以动态设置 tabBar 某一项内容，Object 对象的具体属性如表 4.30 所示。

表 4.30　wx.setTabBarItem 的属性

属性	类型	是否必填	说明
index	number	是	tabBar 的哪一项，从左边算起，0 开始
text	string	是	tab 上的按钮文字
iconPath	string	是	图片路径，icon 大小限制为 40 KB，建议尺寸为 81 px×81 px，当 postion 为 top 时，此参数无效
selectedIconPath	string	是	选中时的图片路径，icon 大小限制为 40 KB，建议尺寸为 81 px×81 px，当 postion 为 top 时，此参数无效
success	Function	否	接口调用成功的回调函数
fail	Function	否	接口调用失败的回调函数
complete	Function	否	接口调用结束的回调函数（调用成功、失败都会执行）

示例代码如下。
```
wx.setTabBarItem({
  index: 0,
  text: 'text',
  iconPath: '/path/to/iconPath',
  selectedIconPath: '/path/to/selectedIconPath'
})
```

5. 在 tabBar 某一项的右上角显示文本

wx.setTabBarBadge（Object object）和 wx.removeTabBarBadge（Object object）是比较常用的功能，可以在 tabBar 右上角显示或隐藏文本，Object 对象有一个 index 属性，从左边算起，index 值从 0 开始算起，根据 index 值来设置哪一个 tabBar 右上角显示或隐藏文本，Object 对象的 text 属性用来显示设置的文本内容。

示例代码如下。
```
onLoad: function () {
  wx.setTabBarBadge({
    index: 0,
    text:'书'
  });
  wx.removeTabBarBadge({
    index: 1
  });
}
```

效果图如图 4.47 所示。

图 4.47　tabBar 右上角显示或隐藏文本

4.5.9　项目实战：任务 11——实现图书搜索功能

1. 任务目标

通过实现图书搜索功能静态布局，学会应用导航组件和导航 API，同时进一步综合应用视图容器组件、基础内容组件、图片组件等组件的页面布局及样式渲染。

图书搜索功能的使用很简单，即从首页单击进入图书搜索页面，在图书搜索页面里，包含搜索框和热门搜索记录，如图 4.48 所示。输入或直接选择搜索关键词，单机"搜索"按钮，得出搜索结果，如图 4.49 所示。

图 4.48　搜索界面及热门搜索

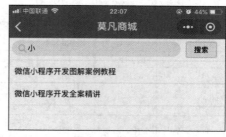

图 4.49　搜索结果

2. 任务实施

下面我们一起来实现图书搜索功能静态布局。

（1）在 app.json 文件里，添加搜索页面 search 页面，具体代码如下。

```
"pages": [
  "pages/index/index",
  "pages/category/category",
  "pages/shoppingcart/shoppingcart",
  "pages/me/me",
  "pages/search/search"
]
```

（2）在首页 index.js 文件里，添加 searchInput 函数来进行 search 搜索页面跳转，在这个函数里使用 wx.navigateTo 将页面跳转到 search 搜索页面，具体代码如下。

```
Page({
  data: {
    indicatorDots: true,
    autoplay: true,
    interval: 5000,
    duration: 1000,
    imgUrls: [
      "/pages/images/haibao/1.jpg",
      "/pages/images/haibao/2.jpg",
      "/pages/images/haibao/3.jpg"
    ],
    hotList: [
      { "id": 1, "listPic": "https://api.mofun365.com:8888/images/goods/1555850845474.jpg", "goodsName": "微信小程序开发图解案例教程", "goodsPrice": 62.8},
      { "id": 2, "listPic": "https://api.mofun365.com:8888/images/goods/1555851154057.jpg", "goodsName": "微信小程序开发全案精讲", "goodsPrice": 41.88 },
      { "id": 3, "listPic": "https://api.mofun365.com:8888/images/goods/1555851345937.jpg", "goodsName": "第一行代码 Java", "goodsPrice": 57.7 }
    ],
    spikeList: [
      { "id": 4, "listPic": "https://api.mofun365.com:8888/images/goods/1555851497575.jpg", "goodsName": "Android 原理解析与开发指南", "goodsPrice": 35.99 },
      { "id": 5, "listPic": "https://api.mofun365.com:8888/images/goods/1555851661073.png", "goodsName": "响应式 Web 开发项目教程", "goodsPrice": 36.4},
      { "id": 6, "listPic": "https://api.mofun365.com:8888/images/goods/1555851817322.jpg", "goodsName": "第一行代码 C 语言", "goodsPrice": 41.99 }
    ],
    bestSellerList: [
      { "id": 7, "listPic": "https://api.mofun365.com:8888/images/goods/1555851965264.jpg", "goodsName": "前端 HTML+CSS 修炼之道", "goodsPrice": 57.7 },
      { "id": 8, "listPic": "https://api.mofun365.com:8888/images/goods/1555850845474.jpg", "goodsName": "微信小程序开发图解案例教程", "goodsPrice": 62.8 },
      { "id": 9, "listPic": "https://api.mofun365.com:8888/images/goods/1555851154057.jpg", "goodsName": "微信小程序开发全案精讲", "goodsPrice": 41.8 }
    ]
  },
  searchInput: function (e) {
    wx.navigateTo({
      url: '../search/search',
    })
  }
})
```

（3）在 search.wxml 文件里进行页面布局，包括搜索输入框布局、热门搜索记录布局、搜索结果布局，具体代码如下。

```
<form bindsubmit="formSubmit" bindreset="formReset">
<view class="search">
  <view class="searchBg">
    <view>
```

```
                <image src="/pages/images/tubiao/search-1.jpg" style="width:20px;height:21px;"></image>
            </view>
            <view>
                <input type="text" placeholder="搜索莫凡商品" placeholder-class="holder" value="{{name}}" name="goodsName"/>
            </view>
        </view>
        <button class="btn" form-type="submit" bindtap="searchGoods">搜索</button>
    </view>
    <view class="hr"></view>
    <block wx:if="{{result.length > 0}}">
        <block wx:for="{{result}}">
            <view class="item" id="{{item.id}}" bindtap='seeDetail'>
                <view class="name">{{item.goodsName}}</view>
                <view class="hr"></view>
            </view>
        </block>
    </block>
    <block wx:else>
        <view class="hotSearch">
            <view class="title">
                <view class="left">热门搜索</view>
                <view class="right" bindtap='refresh'>换一批</view>
            </view>
            <view class="tips">
                <block wx:for="{{goodsNames}}">
                    <view class="tip">{{item.goodsName}}</view>
                </block>
            </view>
        </view>
    </block>
</form>
```

（4）在 search.wxss 文件里对搜索输入框布局、热门搜索记录布局、搜索结果布局进行样式渲染，具体代码如下。

```
.search{
    display: flex;
    flex-direction: row;
    padding:5px;
}
.searchBg{
    background-color: #E8E8ED;
    width:80%;
    border-radius:15px;
    height: 30px;
    display: flex;
    flex-direction: row;
}
.searchBg image{
    margin-left: 10px;
    margin-top: 5px;
}
.search input{
    height: 30px;
    line-height: 30px;
    font-size: 15px;
}
.holder{
    font-size: 13px;
}
.btn{
    font-size: 14px;
    font-weight: bold;
    line-height: 30px;
    margin-left: 10px;
```

```css
        border: 1px solid #cccccc;
        width: 60px;
        text-align: center;
        background-color: #E8E8ED;
        border-radius:3px;
}
.hr{
        border: 1px solid #cccccc;
        opacity: 0.2;
}
.title{
        display: flex;
        flex-direction: row;
        padding: 10px;
}
.left{
        width: 80%;
        font-size: 15px;
}
.right{
        width: 20%;
        font-size: 13px;
        color: #E4393C;
        text-align: right;
}
.tips{
     padding:10px;
     display: flex;
     flex-wrap: wrap;
     justify-content: space-left;
}
.tip{
        background-color: #E8E8ED;
        height:25px;
        line-height: 25px;
        border-radius: 3px;
        text-align: center;
        font-size: 13px;
        margin-right: 10px;
        margin-bottom: 10px;
        padding-left: 5px;
        padding-right: 5px;
}
.item{
        width: 100%;
        padding-left:10px;
        padding-right:10px;
        font-size: 15px;
        padding-top: 10px;
}
.name{
        margin-bottom: 10px;
}
```

（5）在 search.js 文件里对热门搜索记录提供初始化数据，同时提供搜索框，可以输入图书名称进行搜索，具体代码如下。

```js
Page({
  data: {
    result: [ ],
    name: '',
    goodsNames: [
       { "goodsName": "机器学习"},
       { "goodsName": "Spring Cloud"},
```

```
            { "goodsName": "网站" },
            { "goodsName": "Redis" },
            { "goodsName": "Memcached" },
            { "goodsName": "Vue" },
            { "goodsName": "Spring MVC" },
            { "goodsName": "小程序" },
            { "goodsName": "Oracle" },
            { "goodsName": "Java" }
        ]
    },
    formSubmit: function (e) {
        var that = this;
var goodsName = e.detail.value.goodsName;
//演示用数据
        var array = [
            { "id": 1, "goodsName": "微信小程序开发图解案例教程" },
            { "id": 2, "goodsName": "微信小程序开发全案精讲" }
        ];
        that.setData({ result: array });
    }
})
```

这样就完成了搜索输入框布局、热门搜索记录布局和搜索结果布局，搜索结果也是放在 search 页面中，只是用 wx: if 进行了条件判断，如果搜索有结果，就显示结果记录页面，否则显示搜索输入框可以进行搜索。

4.6 项目实战：任务12——实现图书更多列表显示功能静态布局

1. 任务目标

通过实现图书更多列表功能静态布局，综合应用 view 视图容器组件、swiper 滑块视图容器组件、text 文本组件、image 图片组件等，学会使用导航组件和导航 API 进行页面跳转，学会使用 swiper 滑块视图容器组件进行页签的切换显示，学会接收页面路径携带过来的参数。

在图书更多列表页面里，最上面区域是图书搜索区域，中间是热门技术、秒杀时刻、畅销书籍 3 个页签导航，最下面是各个页签导航对应的内容，如图 4.50～图 4.52 所示。

图 4.50　热门技术列表　　　图 4.51　秒杀时刻列表　　　图 4.52　畅销书籍列表

2. 任务实施

下面我们一起来实现图书更多列表显示功能静态布局。

（1）在 app.json 文件里，添加图书更多列表页面 goods 页面，具体代码如下。

```
"pages": [
    "pages/index/index",
    "pages/category/category",
    "pages/shoppingcart/shoppingcart",
    "pages/me/me",
    "pages/search/search",
    "pages/goods/goods"
]
```

（2）在 index.js 文件里，将首页的"查看更多链接"绑定 more 函数，在这个函数里使用 wx.navigateTo 将页面跳转到 goods 图书更多列表页面,它需要携带参数id,用来标识热门技术（id=0）、秒杀时刻（id=1）、畅销书籍分类（id=2），具体代码如下。

```
Page({
    data: {
        indicatorDots: true,
        autoplay: true,
        interval: 5000,
        duration: 1000,
        imgUrls: [
            "/pages/images/haibao/1.jpg",
            "/pages/images/haibao/2.jpg",
            "/pages/images/haibao/3.jpg"
        ],
        hotList: [
            { "id": 1, "listPic": "https://api.mofun365.com:8888/images/goods/1555850845474.jpg", "goodsName": "微信小程序开发图解案例教程", "goodsPrice": 62.8 },
            { "id": 2, "listPic": "https://api.mofun365.com:8888/images/goods/1555851154057.jpg", "goodsName": "微信小程序开发全案精讲", "goodsPrice": 41.88 },
            { "id": 3, "listPic": "https://api.mofun365.com:8888/images/goods/1555851345937.jpg", "goodsName": "第一行代码 Java", "goodsPrice": 57.7 }
        ],
        spikeList: [
            { "id": 4, "listPic": "https://api.mofun365.com:8888/images/goods/1555851497575.jpg", "goodsName": "Android 原理解析与开发指南", "goodsPrice": 35.99 },
            { "id": 5, "listPic": "https://api.mofun365.com:8888/images/goods/1555851661073.png", "goodsName": "响应式 Web 开发项目教程", "goodsPrice": 36.4 },
            { "id": 6, "listPic": "https://api.mofun365.com:8888/images/goods/1555851817322.jpg", "goodsName": "第一行代码 C 语言", "goodsPrice": 41.99 }
        ],
        bestSellerList: [
            { "id": 7, "listPic": "https://api.mofun365.com:8888/images/goods/1555851965264.jpg", "goodsName": "前端 HTML+CSS 修炼之道", "goodsPrice": 57.7 },
            { "id": 8, "listPic": "https://api.mofun365.com:8888/images/goods/1555850845474.jpg", "goodsName": "微信小程序开发图解案例教程", "goodsPrice": 62.8 },
            { "id": 9, "listPic": "https://api.mofun365.com:8888/images/goods/1555851154057.jpg", "goodsName": "微信小程序开发全案精讲", "goodsPrice": 41.8 }
        ]
    },
    searchInput: function (e) {
        wx.navigateTo({
            url: '../search/search',
        })
    },
    more: function (e) {
        var id = e.currentTarget.id;
        wx.navigateTo({
            url: '../goods/goods?id=' + id,
        })
    }
})
```

（3）在 goods.wxml 文件里进行搜索区域、页签导航和页签对应内容的布局设计，具体代码如下。

```xml
<view class="content">
  <view class="search">
    <view class="searchInput" bindtap="searchInput">
      <image src="/pages/images/tubiao/fangdajing-1.jpg" style="width:15px;height:19px;"></image>
      <text class="searchContent">搜索莫凡商品</text>
    </view>
  </view>
  <view class="type">
    <view class="{{currentTab==0?'select':'default'}}" data-current="0" bindtap="switchNav">热门技术</view>
    <view class="{{currentTab==1?'select':'default'}}" data-current="1" bindtap="switchNav">秒杀时刻</view>
    <view class="{{currentTab==2?'select':'default'}}" data-current="2" bindtap="switchNav">畅销书籍</view>
  </view>
  <view class="hr"></view>
  <view>
    <swiper current="{{currentTab}}" style="height:1000px;">
      <swiper-item>
        <view class="list">
          <block wx:for="{{books}}">
            <view class="book" bindtap="seeDetail" id="{{item.id}}">
              <view class="pic">
                <image src="{{item.listPic}}" mode="aspectFit" style="width:115px;height:120px;"></image>
              </view>
              <view class="movie-info">
                <view class="base-info">
                  <view class="name">{{item.goodsName}}</view>
                  <view class="desc">作者:{{item.author}} 著</view>
                  <view class="desc">出版社:{{item.bookConcern}}</view>
                  <view class="desc">出版时间:{{item.publishTime}}</view>
                  <view class="people">
                    <text class="price">¥{{item.goodsPrice}}</text>
                    <text class="org">¥{{item.goodsCost}}</text>
                  </view>
                </view>
              </view>
            </view>
            <view class="hr"></view>
          </block>

        </view>
      </swiper-item>
      <swiper-item>
        <view class="list">
          <block wx:for="{{books}}">
            <view class="book" bindtap="seeDetail" id="{{item.id}}">
              <view class="pic">
                <image src="{{item.listPic}}" mode="aspectFit" style="width:115px;height:120px;"></image>
              </view>
              <view class="movie-info">
                <view class="base-info">
                  <view class="name">{{item.goodsName}}</view>
                  <view class="desc">作者:{{item.author}} 著</view>
                  <view class="desc">出版社:{{item.bookConcern}}</view>
                  <view class="desc">出版时间:{{item.publishTime}}</view>
                  <view class="people">
                    <text class="price">¥{{item.goodsPrice}}</text>
                    <text class="org">¥{{item.goodsCost}}</text>
                  </view>
                </view>
              </view>
            </view>
            <view class="hr"></view>
          </block>

        </view>
```

```
          </swiper-item>
          <swiper-item>
            <view class="list">
              <block wx:for="{{books}}">
                <view class="book" bindtap="seeDetail" id="{{item.id}}">
                  <view class="pic">
                    <image src="{{item.listPic}}" mode="aspectFit" style="width:115px;height:120px;"></image>
                  </view>
                  <view class="movie-info">
                    <view class="base-info">
                      <view class="name">{{item.goodsName}}</view>
                      <view class="desc">作者:{{item.author}} 著</view>
                      <view class="desc">出版社:{{item.bookConcern}}</view>
                      <view class="desc">出版时间:{{item.publishTime}}</view>
                      <view class="people">
                        <text class="price">¥{{item.goodsPrice}}</text>
                        <text class="org">¥{{item.goodsCost}}</text>
                      </view>
                    </view>
                  </view>
                </view>
                <view class="hr"></view>
              </block>
            </view>
          </swiper-item>
        </swiper>
      </View>
</View>
```

（4）在 goods.wxss 文件里对搜索区域、页签导航和页签对应内容进行样式渲染，具体代码如下。

```
.content{
    font-family: "Microsoft YaHei";
    width: 100%;
}
.search{
    width: 100%;
    background-color: #009966;
    height: 50px;
    line-height: 50px;
}
.searchInput{
    width: 95%;
    background-color: #ffffff;
    height: 30px;
    line-height: 30px;
    border-radius: 15px;
    display: flex;
    justify-content:center;
    align-items:center;
    margin: 0 auto;
}
.searchContent{
    font-size:12px;
    color: #777777;
}
.type{
    display: flex;
    flex-direction: row;
    width: 100%;
    margin: 0 auto;
    position: fixed;
    z-index: 999;
    background: #f2f2f2;
    top:50px;
}
.type view{
```

```css
        margin: 0 auto;
}
.select{
    font-size:16px;
    font-weight: bold;
    width: 25%;
    text-align: center;
    height: 45px;
    line-height: 45px;
    border-bottom:5rpx solid #009966;
    color: #009966;
}
.default{
    width: 25%;
    font-size:16px;
    text-align: center;
    height: 45px;
    line-height: 45px;
}
.list{
   margin-top: 50px;
}
.book{
    display: flex;
    flex-direction: row;
    width: 100%;
}
.pic image{
    width:80px;
    height:100px;
    padding:10px;
}
.base-info{
    font-size: 12px;
    padding-top: 10px;
    line-height: 22px;
}
.name{
    font-size: 15px;
    font-weight: bold;
    color: #000000;
}
.people{
    color: #555555;
    margin-top: 5px;
    margin-bottom: 5px;
}
.price{
    font-size: 18px;
    font-weight: bold;
    color: #E53D30;
    margin-left:5px;
}
.org{
   text-decoration: line-through;
   margin-left: 10px;
   margin-right: 5px;
}
.desc{
    color: #333333;
}
.hr{
    height: 1px;
    width: 100%;
    background-color: #009966;
    opacity: 0.2;
```

```
}
.btn{
    position: absolute;
    right: 10px;
    margin-top:50px;
}
.btn button{
    width:52px;
    height: 25px;
    font-size:11px;
    color: red;
    border: 1px solid red;
    background-color: #ffffff;
}
```

（5）在 goods.js 文件里，接收从首页携带过来的参数，需要在 onLoad 生命周期函数里获取图书类别参数，根据图书类别参数来显示对应页签及对应内容，给页签绑定 switchNav 导航切换函数，切换页签的同时切换内容，具体代码如下。

```
Page({
    data: {
        currentTab: 0,  //当前页签对应的序号值
        books: [
            { "id": 1,  "listPic": "https://api.mofun365.com:8888/images/goods/1555850845474.jpg",  "goodsName": "微信小程序开发图解案例教程",  "goodsPrice": 62.8 },
            { "id": 2,  "listPic": "https://api.mofun365.com:8888/images/goods/1555851154057.jpg",  "goodsName": "微信小程序开发全案精讲",  "goodsPrice": 41.88 },
            { "id": 3,  "listPic": "https://api.mofun365.com:8888/images/goods/1555851345937.jpg",  "goodsName": "第一行代码 Java",  "goodsPrice": 57.7 }
        ]
    },
    onLoad: function (e) {
        var type = e.id; //接收携带参数
        console.log(type);
        this.setData({ currentTab: type });//根据携带参数显示对应页签内容
    },
    switchNav: function (e) { //页签导航切换
        var page = this;
        var type = e.target.dataset.current;
        if (this.data.currentTab == type) {
            return false;
        } else {
            page.setData({ currentTab: type });
            this.getBookList(type);
        }
    }
})
```

这样就完成了图书更多列表显示功能静态布局设计。

4.7 小结

本章讲解了视图容器组件，包括 view 视图容器组件、scroll-view 可滚动视图容器组件、swiper 滑块视图容器组件、movable-view 可移动视图容器组件、cover-view 覆盖原生组件的视图容器组件；讲解了基础内容组件，包括 icon 图标组件、text 文本组件、progress 进度条组件、rich-text 富文本组件、editor 富文本编辑器及其 API、image 图片组件及其 API、导航组件及其 API；通过这些组件和 API 的使用来完成页面布局和动态效果。

第5章
莫凡商城首页动态绑定设计

本章讲解莫凡商城首页动态绑定设计，页面通过动态绑定设计可以实现动态交互效果。页面视图与逻辑层进行交互，要用微信小程序函数来处理。微信小程序提供了生命周期函数、页面事件函数、页面路由管理、自定义函数、setData 设值函数。逻辑层的数据渲染到页面视图需要数据动态绑定，在页面视图中借助于双大括号{{}}来取值。微信小程序可以通过网络请求来获取数据、文件、会话等完成动态交互。

5.1 微信小程序函数处理

5.1.1 生命周期函数

在使用 Page()构造器注册页面时，需要使用生命周期函数，包括 onLoad 页面加载时生命周期函数、onShow 页面显示生命周期函数、onReady 页面初次渲染完成生命周期函数、onHide 页面隐藏生命周期函数和 onUnload 页面卸载生命周期函数。

（1）onLoad 页面加载时生命周期函数：一个页面只会调用一次，可以在 onLoad 的参数中获取打开当前页面路径中的参数，接收页面参数可以获取 wx.navigateTo 和 wx.redirectTo 及<navigator/>页面跳转时携带的参数。

（2）onShow 页面显示生命周期函数：每次打开页面都会调用一次，页面显示/切入前台时触发。

（3）onReady 页面初次渲染完成生命周期函数：页面初次渲染完成时触发，一个页面只会调用一次，代表页面已经准备妥当，可以和视图层进行交互，对界面的设置，如 wx.setNavigationBarTitle，需在 onReady 之后进行。

（4）onHide 页面隐藏生命周期函数：页面隐藏/切入后台时触发，如页面之间跳转或通过底部 Tab 切换到其他页面，小程序切入后台等。

（5）onUnload 页面卸载生命周期函数：页面卸载时触发，如页面跳转或者返回到之前的页面时。
示例代码如下。

```
Page({
  onLoad: function (e) {
    console.log("onLoad 页面加载时生命周期函数");
  },
  onShow: function () {
    console.log("onShow 页面显示生命周期函数");
  },
  onReady: function () {
    console.log("onReady 页面初次渲染完成生命周期函数");
  },
  onHide: function () {
    console.log("onHide 页面隐藏生命周期函数");
  },
```

```
onUnload: function () {
    console.log("onUnload 页面卸载生命周期函数");
  },
})
```

页面第一次加载完成时，打印日志如图 5.1 所示。

图 5.1　打印日志

生命周期函数的调用过程如图 5.2 所示。

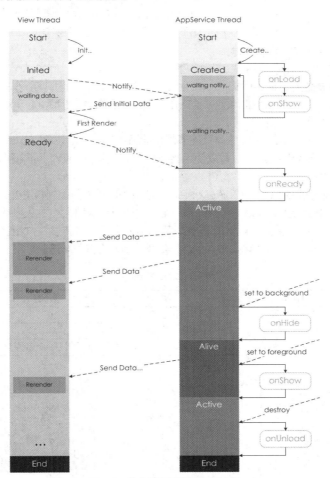

图 5.2　生命周期函数的调用过程

（1）View Thread 代表视图层线程，AppService Thread 代表业务逻辑层线程。

（2）业务逻辑层线程创建、完成时会调用 onLoad 页面加载时生命周期函数、onShow 页面显示生命周期函数。

（3）视图层线程创建完成后，异步通知业务逻辑层线程来获取数据，业务逻辑层线程在给视图层线程发送数据来渲染页面时会调用 onReady 页面初次渲染完成生命周期函数。

（4）页面隐藏/切入后台时会调用 onHide 页面隐藏生命周期函数。

（5）页面显示/切入前台时会调用 onShow 页面显示生命周期函数。

（6）业务逻辑层线程在摧毁时会调用 onUnload 页面卸载生命周期函数。

5.1.2 页面事件函数

微信小程序针对页面事件提供了 5 个函数，分别是 onPullDownRefresh()监听用户下拉刷新事件处理函数、onReachBottom()监听用户上拉触底事件处理函数、onPageScroll（Object object）监听用户滑动页面事件处理函数、onResize()监听页面尺寸发生改变的事件处理函数、onShareAppMessage（Object object）监听用户单击页面内转发处理函数。

（1）onPullDownRefresh()监听用户下拉刷新事件处理函数：需要在 app.json 文件的 window 选项中或页面配置中开启 enablePullDownRefresh。可以通过 wx.startPullDownRefresh 触发下拉刷新，调用后触发下拉刷新动画，效果与用户手动下拉刷新一致。当处理完数据刷新后，wx.stopPullDownRefresh 可以停止当前页面的下拉刷新。

（2）onReachBottom()监听用户上拉触底事件处理函数：可以在 app.json 文件的 window 选项中或页面配置中设置触发距离 onReachBottomDistance。在触发距离内滑动期间，本事件只会被触发一次。

（3）onPageScroll（Object object）监听用户滑动页面事件处理函数：可以获取页面在垂直方向已滚动的距离（单位为像素）。

（4）onResize()监听页面尺寸发生改变的事件处理函数：可以使用页面的 onResize 来监听页面尺寸发生改变的事件。对于自定义组件，可以使用 resize 生命周期来监听。回调函数中将返回显示区域的尺寸信息。

（5）onShareAppMessage（Object object）监听用户单击页面内转发处理函数：监听用户单击页面内转发按钮（button 组件 open-type="share"）或右上角菜单中"转发"按钮的行为，并自定义转发内容，只有定义了此事件处理函数，右上角菜单中才会显示"转发"按钮。此事件需要返回一个 Object 对象，用于自定义分享内容。Object 对象参数说明如表 5.1 所示。

表 5.1 转发参数

字段	说明	默认值
title	分享标题	当前小程序名称
desc	分享描述	当前小程序名称
path	分享路径	当前页面 path，必须是以"/"开头的完整路径

示例代码如下。

```
Page({
  onShareAppMessage: function () {
    return {
      title: '自定义分享标题',
      desc: '自定义分享描述',
      path: '/page/user?id=123'
    }
  }
})
```

5.1.3 页面路由管理

微信小程序的页面路由都是由微信小程序框架来管理的，框架以栈的形式维护了所有页面，栈作为一种数据结构，是一种只能在一端进行插入和删除操作的特殊线性表，它按照后进先出的原则存储数据，先进入的数据被压入栈底，最后进入的数据在栈顶，需要读数据的时候从栈顶开始读出数据（最后进入的一个数据被第一个读出来）。

微信小程序页面交互也是通过栈来完成的。微信小程序初始化时，新页面入栈；打开新页面时，新页面入栈；页面重定向时，当前页面出栈，新页面入栈；页面返回时，页面不断出栈，直到返回指定页面，新页面入栈；Tab（导航标签）切换时，页面全部出栈，只留下新的 Tab 页面；重加载时，页面全部出栈，只留下新的页面。

路由的触发方式及页面生命周期函数如表 5.2 所示。

表 5.2　路由的触发方式及页面生命周期函数

页面路由方式	触发时机	路由后页面	路由前页面
初始化	小程序打开的第一个页面	onLoad, onShow	
打开新页面	调用 API wx.navigateTo 或使用组件 <navigator open-type="navigate"/>	onLoad, onShow	onHide
页面重定向	调用 API wx.redirectTo 或使用组件 <navigator open-type="redirect"/>	onLoad, onShow	onUnload
页面返回	调用 API wx.navigateBack 或用户按左上角的"返回"按钮	onShow	onUnload（多层页面返回每个页面都会按顺序触发 onUnload）
Tab 切换	Function	调用 API wx.switchTab 或使用组件 <navigator open-type="switchTab"/>或用户切换 Tab	

注意：navigateTo、redirectTo 只能打开非 tabBar 页面；switchTab 只能打开 tabBar 页面；reLaunch 可以打开任意页面；页面底部的 tabBar 由页面决定，即只要是定义为 tabBar 的页面，底部都有 tabBar；调用页面路由带的参数可以在目标页面的 onLoad 中获取。

5.1.4　自定义函数

除了初始化数据和生命周期函数外，Page 中还可以定义一些特殊的函数：事件处理函数。在渲染层可以在组件中加入事件绑定，当达到触发事件时，就会执行 Page 中定义的事件处理函数。事件处理函数如表 5.3 所示。

表 5.3　事件处理函数

类型类型	触发条件
bindtap	当用户单击该组件的时候会在该页面对应的 Page 中找到相应的事件处理函数
touchstart	手指触摸动作开始
touchmove	手指触摸后移动
touchcancel	手指触摸动作被打断，如来电提醒、弹窗
touchend	手指触摸动作结束
tap	手指触摸后马上离开
longpress	手指触摸后，超过 350 ms 再离开，如果指定了事件回调函数并触发了这个事件，tap 事件将不被触发
longtap	手指触摸后，超过 350 ms 再离开（推荐使用 longpress 事件代替）
transitionend	会在 WXSS transition 或 wx.createAnimation 动画结束后触发
animationstart	会在一个 WXSS animation 动画开始时触发
animationiteration	会在一个 WXSS animation 一次迭代结束时触发
animationend	会在一个 WXSS animation 动画完成时触发
touchforcechange	支持 3D Touch 的 iPhone 设备，重按时会触发

示例代码如下。
```
<view bindtap="clickMe" id="1" data-hi="WeChat">click me </view>
```

```
Page({
  clickMe:function(e){
    console.log(e);
    console.log("type 代表事件的类型:"+e.type);
    console.log("timeStamp 页面打开到触发事件所经过的毫秒数:" + e.timeStamp);
    console.log("事件源组件的 id:" + e.target.id);
    console.log("事件源组件上由 data-开头的自定义属性组成的集合:" + e.target.dataset.hi);
    console.log("当前组件的 id:" + e.currentTarget.id);
    console.log("当前组件上由 data-开头的自定义属性组成的集合:" + e.currentTarget.dataset.hi);
  }
})
```

打印日志如图 5.3 所示。

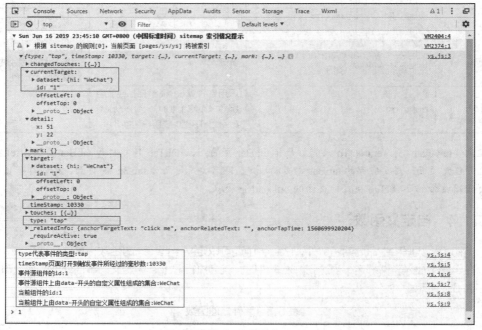

图 5.3　打印日志

5.1.5　setData 设值函数

Page.prototype.setData()为设值函数,用于将数据从逻辑层发送到视图层,同时改变对应的 this.data 的值。

setData()的参数格式:接受一个对象,以 key、value 的形式表示将 this.data 中的 key 对应的值改变成 value。

其中,key 可以非常灵活,以数据路径的形式给出,如 array[2].message, a.b.c.d,并且不需要在 this.data 中预先定义。

示例代码如下。
```
<!--index.wxml-->
<view>{{text}}</view>
<button bindtap="changeText"> Change normal data </button>
```

```
<view>{{array[0].text}}</view>
<button bindtap="changeItemInArray"> Change Array data </button>
<view>{{object.text}}</view>
<button bindtap="changeItemInObject"> Change Object data </button>
<view>{{newField.text}}</view>
<button bindtap="addNewField"> Add new data </button>

//index.js
Page({
  data: {
    text: 'init data',
    array: [{text: 'init data'}],
    object: {
      text: 'init data'
    }
  },
  changeText: function() {
    this.setData({
      text: 'changed data'
    })
  },
  changeItemInArray: function() {
    this.setData({
      'array[0].text':'changed data'
    })
  },
  changeItemInObject: function(){
    this.setData({
      'object.text': 'changed data'
    });
  },
  addNewField: function() {
    this.setData({
      'newField.text': 'new data'
    })
  }
})
```

注意：直接修改 this.data 无效，因为无法改变页面的状态，还会造成数据不一致。单次设置的数据不能超过 1 024 KB，需尽量避免一次设置过多的数据。

5.2 微信小程序网络请求

慕课视频
微信小程序网络请求

微信小程序如果想动态地渲染页面，就需要从后台服务器接口获取数据，不能直接把数据写在页面或者业务逻辑层里。这样的数据都是静态数据，动态数据就需要调用接口发起网络请求来获取，如想通过 https://api.mofun365.com:8888/api/address/getAddressList?provinceId=8 这个省市县接口来获取黑龙江省下面对应市的数据，就可以通过网络请求 API 来向向省市县接口地址发起请求，通过这个 API 返回数据，然后渲染到页面视图上，即可达到动态显示的效果。

5.2.1 网络访问配置

微信小程序在发起网络请求（如 https://api.mofun365.com:8888/api/address/getAddressList?provinceId=8）前，需要在微信公众平台上进行访问域名（如 https://api.mofun365.com:8888，只是域名，无内容）的配置。小程序只允许访问已配置的域名，包括普通 HTTPS 请求（wx.request）、上传文件（wx.uploadFile）、下载文件（wx.downloadFile）和 WebSocket 通信（wx.connectSocket）

的域名。

注意：从基础库 2.4.0 开始，网络请求允许与局域网 IP 通信，但要注意不允许与本机 IP 通信。从基础库 2.7.0 开始，微信小程序提供了 UDP 通信（wx.createUDPSocket），只允许与同个局域网内的非本机 IP 通信。

1. 配置流程

在微信公众平台首页登录，在小程序后台的"开发"→"开发设置"→"服务器域名"中进行配置（需要用注册的微信扫码确认身份），如图 5.4 和图 5.5 所示。

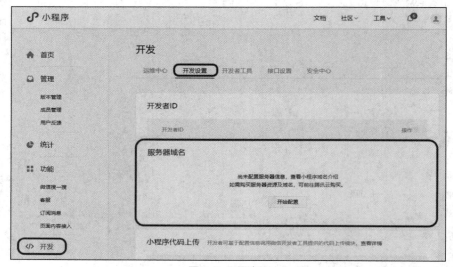

图 5.4　配置流程

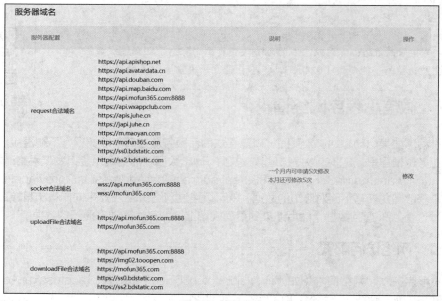

图 5.5　域名配置

如果没有配置域名就直接访问，系统会提示错误信息，如图 5.6 所示。

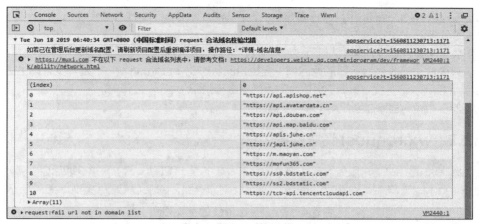

图 5.6 域名访问错误信息

碰到这样的错误信息，有两种可能：一种是没有配置域名；另一种是配置了域名但没有生效，需要刷新项目配置后重新编译项目。

注意（下面的网址只是示例，不是真实网址）：

（1）域名只支持 https（wx.request、wx.uploadFile、wx.downloadFile）和 wss（wx.connect-Socket）协议。

（2）域名不能使用 IP 地址（小程序的局域网 IP 除外）或 localhost。

（3）可以配置端口，如 https: //myserver.com: 8080，但是配置后只能向 https: //myserver.com: 8080 发起请求，如果向 https: //myserver.com、https: //myserver.com: 9091 等 URL 发送请求则会失败。

（4）如果不配置端口，如 https: //myserver.com，那么请求的 URL 中也不能包含端口，即便是默认的 443 端口，如果向 https: //myserver.com: 443 发送请求则会失败。

（5）域名必须经过 ICP 备案。

（6）出于安全考虑，api.weixin.qq.com 不能被配置为服务器域名，相关 API 也不能在小程序内调用。开发者应将 AppSecret 保存到后台服务器中，通过服务器使用 getAccessToken 接口获取 access_token，并调用相关 API。

（7）对于每个接口，分别可以配置最多 20 个域名。

（8）网络请求默认超时时间和最大超时时间都是 60 s，超时时间可以在 app.json 文件或 game.json 文件中通过 networktimeout 配置。

2. 使用限制

网络访问也有使用限制，包括网络请求设置、并发限制设置、超时设置、编码设置等使用上的限制。

（1）网络请求的请求来源 referer header 不可设置。其格式固定为 https: //servicewechat.com/{appid}/{version}/page-frame.html，其中{appid}为小程序的 appid，{version}为小程序的版本号，版本号为 0 表示为开发版本、体验版本及审核版本，版本号为 devtools 表示为开发者工具，其余为正式版本号。

（2）wx.request、wx.uploadFile 和 wx.downloadFile 的最大并发限制是 10 个。

（3）wx.connectSockt 的最大并发限制是 5 个。

（4）小程序进入后台运行后，如果 5 s 内网络请求没有结束，会回调错误信息 fail interrupted；在回到前台之前，网络请求接口都会无法调用。

（5）建议服务器返回值使用 UTF-8 编码。对于非 UTF-8 编码，小程序会尝试进行转换，但会有转换失败的可能。

（6）小程序会自动对 BOM 头进行过滤（只过滤一个 BOM 头）。

（7）只要成功接收到服务器的返回值，无论返回状态码（statusCode）是多少，都会进入成功（success）回调。请开发者根据业务逻辑对返回值进行判断。

下面详细讲解小程序的网络请求 API 及其用法。

5.2.2 wx.request 请求数据 API

wx.request 是用来请求服务器数据的 API，它发起的是 HTTPS 请求，来获取后台服务器接口的数据。wx.request（Object object）有一个 Object 对象参数，Object 参数说明如表 5.4 所示。

表 5.4 wx.Request 的参数说明

属性	类型	是否必填	说明
url	String	是	开发者服务器接口地址
data	string/object/ArrayBuffer	否	请求的参数
header	Object	否	设置请求的请求头 header，请求头 header 中不能设置请求来源 Referer，内容类型 content-type 默认为 application/json
method	String	否	请求方法，默认为 GET，有效值为 OPTIONS、GET、HEAD、POST、PUT、DELETE、TRACE、CONNECT
dataType	String	否	数据类型，默认为 json，如果设置 dataType 为 json，系统会尝试对响应的数据做一次 JSON 数据解析 JSON.parse，设置为其他值则不对返回的内容进行 JSON 数据解析 JSON.parse
responseType	String	text	响应的数据类型，text 表示响应的数据为文本，arraybuffer 表示响应的数据为 ArrayBuffer
success	Function	否	收到开发者服务成功返回的回调函数，res = {data: '开发者服务器返回的内容'}
fail	function	否	接口调用失败的回调函数
complete	function	否	接口调用结束的回调函数（调用成功、失败都会执行）

发起 wx.request 请求时，系统也创建了 RequestTask 对象，这个对象提供了以下 3 种方法。

（1）RequestTask.abort() 中断请求任务。

（2）RequestTask.onHeadersReceived（function callback）监听 HTTP Response Header 事件。

（3）RequestTask.offHeadersReceived（function callback）取消监听 HTTP Response Header 事件。

下面演示 wx.request 请求服务器数据 API 的使用方法。

在 JS 文件的 onLoad 函数里，使用 wx.request 请求服务器数据，具体代码如下。

```
Page({
  onLoad:function(){
    var requestTask = wx.request({
      url:'https://api.mofun365.com:8888/api/address/getAddressList',
      data:{
        provinceId:'8'
      },
      method:'GET',
      success:function (res) {
        console.log(res);
      },
```

```
      fail:function(){
        // fail
      },
      complete:function(){
        // complete
      }
    });
    //监听 HTTP Response Header 事件
    requestTask.onHeadersReceived(function(res){
      console.log("-----------监听 HTTP Response Header 事件-------------");
      console.log(res);
    });
    //取消监听 HTTP Response Header 事件
    requestTask.offHeadersReceived(function(){
      console.log("-----------取消监听 HTTP Response Header-------------");
    });
    //中断请求任务
    //requestTask.abort();

  }
})
```

data 数据说明最终发送给服务器的数据是 string 类型的，传入的 data 如果不是 string 类型的，会被转换成 string，转换规则如下。

（1）对于 header['content-type']为'Application/json'的数据，会对数据进行 JSON 序列化。

（2）对于 header['content-type']为'Application/x-www-form-urlencoded'的数据，会将数据转换成 query string(encodeURIComponent(k)=encodeURIComponent(v)&encodeURIComponent(k)=encodeURIComponent(v)...)。

服务器返回数据如图 5.7 所示。

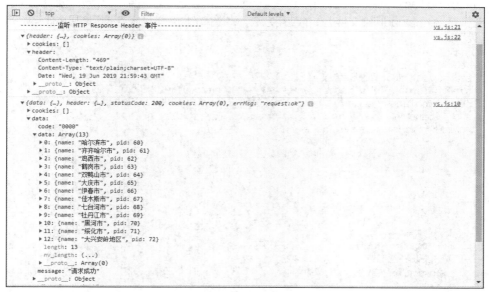

图 5.7　服务器返回数据

requestTask 是 wx.request 创建的一个对象，它可以使用 requestTask.abort()中断请求任务，停止发起网络请求；可以使用 requestTask.onHeadersReceived（function callback）监听 HTTP

Response Header 事件，从打印日志可以看出这个事件会比请求完成事件更早完成；使用 requestTask.offHeadersReceived（function callback）取消监听 HTTP Response Header 事件。

5.2.3　wx.uploadFile 文件上传 API

wx.uploadFile API 可以将本地资源上传到服务器。客户端发起一个 HTTPS POST 请求，其中 content-type 为 multipart/form-data，wx.uploadFile（object）的参数说明如表 5.5 所示。

表 5.5　wx.uploadFile 的参数说明

属性	类型	是否必填	说明
url	string	是	开发者服务器 URL
filePath	string	是	要上传文件资源的路径
name	string	是	文件对应的 key，开发者在服务端通过这个 key 可以获取到文件的二进制内容
header	Object	否	HTTP 请求头 header，请求头 header 中不能设置请求来源 Referer
formData	Object	否	HTTP 请求中其他额外的请求参数
success	Function	否	接口调用成功的回调函数
fail	Function	否	接口调用失败的回调函数
complete	Function	否	接口调用结束的回调函数（调用成功、失败都会执行）

发起 wx.uploadFile 请求时，系统也创建了 UploadTask 对象，这个对象提供了以下 5 种方法。

（1）UploadTask.abort()中断请求任务。
（2）UploadTask.onHeadersReceived（function callback）监听 HTTP Response Header 事件。
（3）UploadTask.offHeadersReceived（function callback）取消监听 HTTP Response Header 事件。
（4）UploadTask.onProgressUpdate（function callback）监听上传进度变化事件。
（5）UploadTask.offProgressUpdate（function callback）取消监听上传进度变化事件。

下面演示 wx.uploadFile 文件上传的使用方法，将选择的图片传到服务器里，具体代码如下。

```
Page({
  onLoad:function(){
    wx.chooseImage({
      count:9, //最多可以选择的图片张数，默认为9
      sizeType:['original', 'compressed'], //original 原图，compressed 压缩图，默认二者都有
      sourceType:['album', 'camera'], // album 从相册选图，camera 使用相机，默认二者都有
      success:function(res){
        var tempFilePaths=res.tempFilePaths;
        const uploadTask=wx.uploadFile({
          url: 'https://api.mofun365.com:8888/api/banner/wxUploadFile',
          filePath:tempFilePaths[0],
          name:'file',
          header:{
            'content-type':'Application/json'
          },
          formData:{
            imgName:'我是图片名称',
            imgSize:'122kb',
            position:'wx' //自定义文件存放的文件夹
          },
          success:function (res) {
            console.log(res);
          }
        });
        //监听 HTTP Response Header 事件
```

```
            uploadTask.onHeadersReceived(function (res) {
                console.log("----------监听 HTTP Response Header 事件--------------");
                console.log(res);
            });
            //取消监听 HTTP Response Header 事件
            uploadTask.offHeadersReceived(function () {
                console.log("----------取消监听 HTTP Response Header--------------");
            });
            //监听上传进度变化事件
            uploadTask.onProgressUpdate(function (res) {
                console.log("----------监听上传进度变化事件--------------");
                console.log(res);
            });
            //取消监听上传进度变化事件
            uploadTask. offProgressUpdate (function () {
                console.log("----------取消监听上传进度变化事件--------------");
            });
            //中断请求任务
            //uploadTask.abort();

        }

    })
  }
})
```

打印日志如图 5.8 所示。

图 5.8 打印日志

uploadTask 是 wx.uploadFile 创建的一个对象，它可以使用 uploadTask.abort()中断请求任务，停止发起网络请求；使用 uploadTask.onHeadersReceived（function callback）监听 HTTP Response Header 事件，从打印日志可以看出这个事件会比请求完成事件更早完成；使用 uploadTask.offHeadersReceived（function callback）取消监听 HTTP Response Header 事件；使用 uploadTask.onProgressUpdate（function callback）监听上传进度变化事件；使用 uploadTask.offProgressUpdate（function callback）

取消监听上传进度变化事件。

5.2.4 wx.downloadFile 文件下载 API

wx.uploadFile 是文件上传的 API，wx.downloadFile 是文件下载的 API，它们正好相反。wx.downloadFile 是客户端直接发起一个 HTTPS GET 请求，从服务器获得数据，返回文件的本地临时路径，单次下载允许的最大文件大小为 50 MB，下载到微信小程序客户端本地，参数说明如表 5.6 所示。

表 5.6 wx.downloadFile 的参数说明

属性	类型	是否必填	说明
url	string	是	开发者服务器 URL
header	Object	否	HTTP 请求头 header，请求头 header 中不能设置请求来源 Referer
filePath	string	否	指定文件下载后存储的路径
success	Function	否	收到开发者服务器成功返回的回调函数，res={data: '开发者服务器返回的内容'}
fail	Function	否	接口调用失败的回调函数
complete	Function	否	接口调用结束的回调函数（调用成功、失败都会执行）

DownloadTask 对象是 wx.downloadFile 创建的一个对象，它可以监听下载进度变化事件，以及取消下载任务的对象，DownloadTask 对象提供了以下方法。

（1）DownloadTask.abort()中断下载任务。

（2）DownloadTask.onProgressUpdate（function callback）监听下载进度变化事件。

（3）DownloadTask.offProgressUpdate（function callback）取消监听下载进度变化事件。

（4）DownloadTask.onHeadersReceived（function callback）监听 HTTP Response Header 事件，会比请求完成事件更早。

（5）DownloadTask.offHeadersReceived（function callback）取消监听 HTTP Response Header 事件。

下面演示一下 wx.downloadFile 文件下载接口的使用。从服务器获取一张图片，然后将其下载到微信小程序客户端，显示出来。

（1）在 WXML 文件里，添加 image 组件，用来显示服务器传递过来的图片，具体代码如下。

```
<image src="{{src}}" style="width:270px;height:126px;"></image>
```

（2）在 JS 文件里，下载服务器中的一张图片，将它的临时路径赋值给 src，具体代码如下。

```
Page({
  data: {
    src: ''
  },
  onLoad:function() {
    var page=this;
    const downloadTask=wx.downloadFile({
      url: "https://api.mofun365.com:8888/images/banner/1555848473813.jpg",
      type: 'image',   // 下载资源的类型,用于客户端识别处理,有效值:image/audio/video
      success:function(res) {
        console.log(res);
        var tempPath=res.tempFilePath;
        page.setData({
          src: tempPath
        });
      }
    });
    //监听 HTTP Response Header 事件
```

```
        downloadTask.onHeadersReceived(function(res) {
            console.log("----------监听 HTTP Response Header 事件--------------");
            console.log(res);
        });
        //取消监听 HTTP Response Header 事件
        downloadTask.offHeadersReceived(function() {
            console.log("----------取消监听 HTTP Response Header--------------");
        });
        //监听下载进度变化事件
        downloadTask.onProgressUpdate(function(res) {
            console.log("----------监听下载进度变化事件-------------");
            console.log(res);
        });
        //取消监听下载进度变化事件
        downloadTask.offProgressUpdate(function() {
            console.log("----------取消监听下载进度变化事件-------------");
        });
        //中断请求任务
        //downloadTask.abort();
    }
})
```

打印日志如图 5.9 和图 5.10 所示。

图 5.9 监听变化

图 5.10 返回值

从打印日志里可以看出，首先执行的是 DownloadTask.onHeadersReceived（function callback）监听 HTTP Response Header 事件，它比请求完成事件更早执行；然后执行的是 DownloadTask.onProgressUpdate（function callback）监听下载进度变化事件，直到下载完成；最后返回文件的临时路径，根据临时路径就可以将文件渲染到视图里或者下载到手机客户端。

5.2.5 WebSocket 会话 API

WebSocket 会话 API 用来创建一个会话连接，创建完会话连接后可以相互通信，像微信聊天和 QQ 聊天一样。它会用到以下 7 种方法。

（1）wx.connectSocket（OBJECT）创建一个会话连接。

（2）wx.onSocketOpen（CALLBACK）监听 WebSocket 连接打开事件。

（3）wx.onSocketError（CALLBACK）监听 WebSocket 错误。
（4）wx.sendSocketMessage（OBJECT）发送数据。
（5）wx.onSocketMessage（CALLBACK）监听 WebSocket 接收到服务器的消息事件。
（6）wx.closeSocket()关闭 WebSocket 连接。
（7）wx.onSocketClose（CALLBACK）监听 WebSocket 关闭。
wx.connectSocket（object）的参数说明如表 5.7 所示。

表 5.7 wx.connectSocket 的参数说明

属性	类型	是否必填	说明
url	string	是	开发者服务器 URL
data	Object	否	请求的数据
header	Object	否	HTTP 请求头 header，请求头 header 中不能设置请求来源 Referer
method	string	否	请求方法，默认是 GET，有效值为 OPTIONS、GET、HEAD、POST、PUT、DELETE、TRACE、CONNECT
success	Function	否	接口调用成功的回调函数
fail	Function	否	接口调用失败的回调函数
complete	Function	否	接口调用结束的回调函数（调用成功、失败都会执行）

一个微信小程序同时只能有一个 WebSocket 连接，如果创建时已存在一个 WebSocket 连接，则会自动关闭该连接，并重新创建一个 WebSocket 连接。

wx.sendSocketMessage（object）的参数说明如表 5.8 所示。

表 5.8 wx.sendSocketMessage 的参数说明

属性	类型	是否必填	说明
data	string/ArrayBuffer	否	请求的数据
success	Function	否	接口调用成功的回调函数
fail	Function	否	接口调用失败的回调函数
complete	Function	否	接口调用结束的回调函数（调用成功、失败都会执行）

下面通过 WebSocket 会话来实现聊天功能。

（1）在 WXML 文件里进行界面布局设计，默认微信聊天界面设计，来实现聊天功能，具体代码如下。

```
<view>
    <block wx:for="{{resData}}" wx:for-item="item2">
        <view style="font-weight:bold;margin:10px;">
            <text>服务端</text>
            <text style="font-size:15px;color:green;border:1px solid #cccccc;border-redis:5px;margin-left:10px;padding:5px;">{{item2}}</text>
        </view>
    </block>
</view>

<view style="margin-top:20px;">
    <block wx:for="{{sendMsg}}" wx:for-item="item1">
        <view style="font-weight:bold;margin:10px;text-align:right;">
            <text style="font-size:15px;color:red;border:1px solid #cccccc;border-redis:5px;margin-left:10px;padding:5px;">{{item1}}</text>
            <text style="margin-left:10px;">客户端</text>
        </view>
    </block>
</view>
```

```html
<view style="position:fixed;bottom:0px">
  <view style="display:flex;flex-direction:row;margin:10px;">
    <input type="text" name="msg" bindblur="getMsg" value="{{info}}" style="width:200px;border:1px solid #cccccc;" />
    <button type="primary" size="mini" bindtap="send" style="margin-left:10px">发送</button>
    <button type="primary" size="mini" bindtap="closeConn" style="margin-left:10px">关闭</button>
  </view>
</view>
```

（2）在 JS 文件里，利用 WebSocket 的 API 创建一个会话连接，练习监听连接打开成功和失败、发送信息并接收信息、关闭连接等 API 的使用，具体代码如下。

```javascript
Page({
  data:{
    msg:'', //输入框发送信息
    sendMsg:[], //客户端发送信息集合
    socketOpen:false, //开启 WebSockect
    resData:[], //服务端接收信息集合
    info:'' //输入框默认信息
  },
  onLoad: function(){
    this.createConn();//页面加载时打开 WebSocket 连接
  },
  createConn: function(){//创建 WebSocket 连接
    var page = this;
    wx.connectSocket({
      url: 'wss://mofun365.com/api/socketServer',
      data: {
        x: '',
        y: ''
      },
      header:{
        'content-type': 'Application/json'
      },
      method:"GET"
    });
    wx.onSocketOpen(function(res) {//监听 WebSocket 连接是否打开
      console.log(res);
      page.setData({
        socketOpen: true
      });
      console.log('WebSocket 连接已打开！')
    });
    wx.onSocketError(function(res) {//监听 WebSocket 连接失败
      console.log('WebSocket 连接打开失败，请检查！')
    })
  },
  send: function(e) {//发送 WebSocket 会话信息
    if (this.data.socketOpen) {
      console.log(this.data.socketOpen);
      wx.sendSocketMessage({
        data: this.data.msg
      });
      var sendMsg = this.data.sendMsg;
      sendMsg.push(this.data.msg);
      this.setData({
        sendMsg:sendMsg
      });
      this.setData({info:''});
      var page = this;
      wx.onSocketMessage(function(res) {//监听 WebSocket 信息，接收服务器返回内容
        var resData = page.data.resData;
        resData.push(res.data);
```

```
            page.setData({
                resData:resData
            });
            console.log(resData);
            console.log('收到服务器内容:' + res.data)
        })
    } else {
        console.log('WebSocket 连接打开失败,请检查!');
    }
},
closeConn: function(e) {//关闭 WebSocket 连接
    wx.closeSocket();
    wx.onSocketClose(function(res) {
        console.log('WebSocket 已关闭!')
    });
},
getMsg: function(e) {//获取输入框信息
    var page = this;
    page.setData({
        msg:e.detail.value
    });
}
})
```

（3）微信小程序客户端和服务端代码写完之后，页面加载时开始创建 WebSocket 连接，同时可以发送信息给服务端以及接收服务端传递过来的信息，通过输入框发送信息，单击"发送"按钮将信息发送出去，单击"关闭"按钮关闭会话连接，如图 5.11 所示。

图 5.11　会话聊天

5.2.6　项目实战：任务 10（2）——实现图书列表显示功能动态渲染

1. 任务目标

通过实现图书列表显示功能动态渲染，学会用微信小程序发起网络请求及将数据渲染到页面上，实现图书列表动态渲染的效果。

图书列表包括 3 个区域：热门技术区域、秒杀时刻区域、畅销书籍区域，这 3 个区域的布局方式一样，都是最多显示 3 本书，如图 4.42 和图 4.43 所示。

2. 任务实施

在 4.4.3 小节中，我们已完成静态页面布局设计，下面一起来实现动态获取这些书籍列表和动态渲染页面效果。

（1）在 index.js 文件里添加发起网络请求，编写获取图书列表的 getBookList 方法，具体代码如下。

```javascript
var app = getApp();
var host = app.globalData.host;
Page({
  data: {
    indicatorDots:true,
    autoplay: true,
    interval: 5000,
    duration: 1000,
    imgUrls: [
      "/pages/images/haibao/1.jpg",
      "/pages/images/haibao/2.jpg",
      "/pages/images/haibao/3.jpg"
    ],
    hotList:[], //热门书籍
    spikeList:[], //秒杀时刻
    bestSellerList:[], //畅销书籍
    host: host
  },
  onLoad:function(options){
    var page = this;
    page.getBookList();
  },
  getBookList:function(){//获取图书列表方法
    var page = this;
    wx.request({
      url:host + '/api/goods/getHomeGoodsList',
      method:'GET',
      data:{},
      header:{
        'Content-Type': 'application/json'
      },
      success:function(res){
        var book=res.data.data;
        var hotList=book.rmjs;//热门书籍列表
        var spikeList=book.mssk;//秒杀时刻列表
        var bestSellerList =book.cxsj;//畅销书籍列表
        page.setData({ hotList: hotList });
        page.setData({ spikeList: spikeList });
        page.setData({ bestSellerList: bestSellerList });
      }
    })
  },
  more:function(e){
    var id = e.currentTarget.id;
    wx.navigateTo({
      url:'../goods/goods?id='+id,
    })
  },
  searchInput:function(e){
    wx.navigateTo({
      url:'../search/search',
    })
```

```
    }
})
```

（2）通过 getBookList 获取图书列表的方法可以动态获取热门书籍列表、秒杀时刻图书列表、畅销书籍列表，查看页面可以看到页面进行了动态渲染，如图 4.42 所示。

5.3 微信小程序定义模板

微信小程序在 WXML 页面上提供模板（template）功能，是为了解决一些重复代码和布局设计的问题，把这些重复的代码抽取出来，放在模板里，当用到相同功能和布局页面时，就可以达到复用的效果。例如，导航菜单、版权信息等各个页面都共用的功能，就可以放在模板里。

5.3.1 定义模板

使用<template/>标签来定义模板代码片段，使用 template 模板的 name 属性作为模板的名字，页面在调用模板的时候，可以根据名字找到相应的模板，同时模板也可以接收传递过来的数据，如{{index}}、{{msg}}、{{time}}都是接收的数据，示例代码如下。

```
<template name="msgItem">
  <view>
    <text> {{index}}: {{msg}} </text>
    <text> Time: {{time}} </text>
  </view>
</template>
```

5.3.2 使用模板

在 WXML 页面文件里，使用 is 属性找到要引入的模板名称，如 msgItem 就是模板的名称，使用 data 属性来传递模板所需要的数据，示例代码如下。

```
<!-- wxml -->
<template is="msgItem" data="{{item}}"/>
```

```
<!-- js -->
Page({
  data: {
    item: {
      index: 100,
      msg: '我是一个模板',
      time: '2019-09-15'
    }
  }
})
```

is 属性可以使用三元运算语法，从而动态决定具体需要渲染哪个模板。下面定义两套模板，奇数使用第一套模板，偶数使用第二套模板。

```
<template name="first">
  <view> 我是第一套模板 </view>
</template>
<template name="second">
  <view> 我是第二套模板 </view>
</template>

<block wx:for="{{[1, 2, 3, 4, 5]}}">
    <template is="{{item % 2 == 0 ? 'first' : 'second'}}"/>
</block>
```

5.4 微信小程序的引用功能

微信小程序的引用有两种方式：import 引用和 include 引用。import 是引用模板文件，在 WXML 页面文件定义模板后，就可以使用 import 将模板引入页面；include 是直接将 WXML 页面内容引入到另一个页面里，但是它不能引用<template/>模板内容。

5.4.1 import 引用

import 可以将<template/>模板引入到页面中使用。

假如在 temp.wxml 文件中定义了一个叫 msg 的模板，引用的示例代码如下。

```
<!-- temp.wxml -->
<template name="msg">
    <text>我是模板内容</text>
<text>{{text}}</text>
</template>
```

在 index.wxml 文件中引用了 temp.wxml 文件，就可以使用模板了，示例代码如下。

```
<import src="temp.wxml"/>
<template is="msg" data="{{text: '你好'}}"/>
```

注意 import 作用域的问题，假如 C import B，B import A，则在 C 中可以使用 B 定义的模板，在 B 中可以使用 A 定义的模板，但是 C 不能使用 A 定义的模板。

5.4.2 include 引用

include 可以将 WXML 页面文件的整个代码引入到目标文件里，但是不能引入<template/>模板文件，相当于是将 WXML 文件拷贝到 include 位置，示例代码如下。

```
<!-- index.wxml -->
<include src="header.wxml"/>
<view> body </view>
<include src="footer.wxml"/>
```

```
<!-- header.wxml -->
<view> 我是头部信息 </view>
```

```
<!-- footer.wxml -->
<view> 我是版权信息 </view>
```

5.5 WXS 小程序脚本语言

WXS（WeiXin Script）是小程序的一套脚本语言，结合 WXML 页面文件，可以构建出页面的结构。它是把原来放在 JS 文件里进行处理的逻辑，直接放在 WXML 页面文件里进行处理。它有两种使用方式：一种是将 WXS 脚本语言嵌入到 WXML 页面文件里，用来在 WXML 文件中的<wxs>标签内处理相关逻辑；另一种是作为以.wxs 为后缀名的文件独立存在，然后再引入到 WXML 页面文件里使用。

1. 嵌入 WXML 页面文件

嵌入 WXML 页面文件的使用方法示例代码如下。

```
<!--wxml-->
<wxs module="m1">
var msg = "hello world";

module.exports.message = msg;
```

```
</wxs>
<view> {{m1.message}} </view>
```

2. 独立为 WXS 文件

在指定的项目目录里面，单击鼠标右键，在弹出的快捷菜单中选择"新建 WXS"可以创建.wxs 文件，如图 5.12 所示。

图 5.12 创建.wxs 脚本语言文件

示例代码如下。

```
// /pages/tools.wxs
var foo = "'hello world' from tools.wxs";
var bar = function(d) {
  return d;
}
module.exports = {
  FOO: foo,
  bar: bar,
};
module.exports.msg = "some msg";
```

```
<!-- page/index/index.wxml -->
<wxs src="./../tools.wxs" module="tools" />
<view> {{tools.msg}} </view>
<view> {{tools.bar(tools.FOO)}} </view>
```

5.5.1 模块化

WXS 代码无论编写在 WXML 文件中的 <wxs> 标签内还是以.wxs 为后缀名的文件内，都是以单独的模块形式存在的。在一个模块中定义的变量与函数，默认为私有，对其他模块不可见。一个模块要想对外暴露其内部的私有变量与函数，只能通过 module.exports 来实现，示例代码如下。

```
// /pages/comm.wxs
var foo = "'hello world' from comm.wxs";
var bar = function(d) {
  return d;
}
module.exports = {
  foo: foo,
  bar: bar
};
```

在.wxs 模块中引用其他 WXS 文件模块，可以使用 require 函数。在 WXS 文件里只能引用.wxs 文件模块，且必须使用相对路径。.wxs 文件模块均为单例，多个页面、多个地方、多次引用，使用的都是同一个.wxs 文件模块对象，如果一个.wxs 文件模块在定义之后，一直没有被引用，则该模块不会被解析与运行，示例代码如下。

```
// /pages/tools.wxs
var foo = "'hello world' from tools.wxs";
var bar = function (d) {
  return d;
}
module.exports = {
  FOO: foo,
  bar: bar,
};
module.exports.msg = "some msg";


// /pages/logic.wxs
var tools = require("./tools.wxs");
console.log(tools.FOO);
console.log(tools.bar("logic.wxs"));
console.log(tools.msg);
```

```
<!-- /page/index/index.wxml -->
<wxs src="./../logic.wxs" module="logic" />
```

5.5.2 变量与数据类型

1. 变量的使用

WXS 中的变量均为值的引用，如果只声明变量而不赋值，则默认值为 undefined，示例代码如下。

```
var foo = 1;
var bar = "hello world";
var i; // i === undefined
```

变量名的命名规则如下。

（1）首字符必须是字母（a~z，A~Z）、下划线（_）。

（2）剩余字符可以是字母（a~z，A~Z）、下划线（_）、数字（0~9）。

（3）保留标识符 delete、void、typeof、null、undefined、NaN、Infinity、var、if、else、true、false、require、this、function、arguments、return、for、while、do、break、continue、switch、case、default 不能作为变量名。

2. 数据类型

WXS 小程序脚本语言支持的数据类型为 number（数值类型）、string（字符串类型）、boolean（布尔值类型）、object（对象类型）、function（函数类型）、array（数组类型）、date（日期类型）、regexp（正则类型）。

（1）number 数值类型包括两种数值：整数和小数，用法如下。

```
var a = 10;
var PI = 3.141592653589793;
```

（2）string 字符串类型有以下两种写法。

```
'hello world';
"hello world";
```

（3）boolean 布尔值类型只有两个特定的值：true 和 false。

（4）object 对象类型是一种无序的键值对，使用方法如下。

```
var o = {}  //生成一个新的空对象
//生成一个新的非空对象
o = {
  'string'   : 1,     //object 的 key 可以是字符串
  const_var : 2,      //object 的 key 也可以是符合变量定义规则的标识符
  func      : {},     //object 的 value 可以是任何类型
};
//对象属性的读操作
console.log(1 === o['string']);
console.log(2 === o.const_var);

//对象属性的写操作
o['string']++;
o['string'] += 10;
o.const_var++;
o.const_var += 10;

//对象属性的读操作
console.log(12 === o['string']);
console.log(13 === o.const_var);
```

（5）function 函数类型支持以下定义方式。

```
//方法 1
function a (x) {
   return x;
}

//方法 2
var b = function (x) {
   return x;
}
```

function 同时也支持以下语法（匿名函数、闭包等）。

```
var a = function (x) {
   return function () { return x;}
}
```

（6）array 数组类型支持以下定义方式。

```
var a = [];          //生成一个新的空数组
a = [1, "2", {}, function(){}];  //生成一个新的非空数组，数组元素可以是任何类型
```

（7）date 日期类型生成 date 对象需要使用 getDate 函数，返回一个当前时间的对象，示例代码如下。

```
getDate()
getDate(milliseconds)
getDate(datestring)
getDate(year, month[, date[, hours[, minutes[, seconds[, milliseconds]]]]])
参数：
milliseconds: 从 1970 年 1 月 1 日 00:00:00 UTC 开始计算的毫秒数
datestring: 日期字符串，其格式为:"month day, year hours:minutes:seconds"
```

（8）regexp 正则类型生成 regexp 对象需要使用 getRegExp 函数，示例代码如下。

```
getRegExp(pattern[, flags])
参数：
pattern: 正则表达式的内容。
flags:修饰符。该字段只能包含以下字符：
  g: global
  i: ignoreCase
  m: multiline。
```

5.5.3 注释

WXS 小程序脚本语言注释有 3 种方式：单行注释、多行注释、结尾注释，示例代码如下。

```
<wxs module="sample">
// 方法一:单行注释
//var name = "小刚";

//方法二:多行注释
```

```
/*
var a = 1;
var b = 2;
*/

//方法三:结尾注释。即从 /* 开始往后的所有 WXS 代码均被注释
/*
var a = 1;
var b = 2;
var c = "fake";

</wxs>
```

5.5.4 语句

WXS 微信小程序脚本语言里,可以使用 if 条件语句、switch 条件语句、for 循环语句和 while 循环语句。

1. if 语句

在 WXS 中,可以使用以下格式的 if 语句:if...else if...else statementN。通过该句型,可以在 statement1~statementN 中选择其中一个执行,示例语法如下。

```
if (表达式) {
  代码块;
} else if (表达式) {
  代码块;
} else if (表达式) {
  代码块;
} else {
  代码块;
}

var age = 10;
if(age < 18){
  console.log("未成年");
}else if(age < 28){
  console.log("青年");
}else{
  console.log("壮年");
}
```

2. switch 语句

switch 语句根据表达式的值与 case 变量值做比较,哪个 case 变量值与表达式的值相等就执行哪个 case 语句,default 分支可以省略不写,case 关键词后面只能使用变量、数字和字符串。如果不写 break 结束语句,程序就会向下继续执行其他满足条件的 case 语句,示例语法如下。

```
switch (表达式) {
  case 变量:
    语句;
  case 数字:
    语句;
    break;
  case 字符串:
    语句;
  default:
    语句;
}

var exp = 10;
switch ( exp ) {
case "10":
  console.log("string 10");
```

```
    break;
case 10:
    console.log("number 10");
    break;
case exp:
    console.log("var exp");
    break;
default:
    console.log("default");
}
```

3. for 循环语句

for 循环语句用来遍历集合,支持使用 break、continue 关键词,示例语法如下。

```
for (语句; 语句; 语句) {
    代码块;
}
for (var i = 0; i < 3; ++i) {
    console.log(i);
    if( i >= 1) break;
}
```

4. while 循环语句

while 循环语句当表达式为 true 时,循环执行语句或代码块,支持使用 break、continue 关键词,示例语法如下。

```
while (表达式){
    代码块;
}
do {
    代码块;
} while (表达式)
```

5.6 下拉刷新及窗口设置

慕课视频

下拉刷新及窗口设置

页面下拉刷新是经常会用到的一个功能,有了这个功能,我们就可以通过刷新页面来获取更多的数据。微信小程序也提供了开始下拉刷新、停止下拉刷新及动态设置背景色等功能。要实现下拉刷新效果,就需要在小程序公共设置 app.json 文件里或者在各个页面的.json 文件里配置 enablePullDownRefresh=true,提供两个事件:onPullDownRefresh()用户下拉刷新事件、onReachBottom()用户上拉触底事件;提供两个 API 接口:wx.startPullDownRefresh 开始下拉刷新、wx.stopPullDownRefresh 停止当前页面下拉刷新。

5.6.1 下拉刷新 API 及事件

1. wx.startPullDownRefresh 开始下拉刷新

微信小程序使用 wx.startPullDownRefresh(Object object)来进行刷新,调用后触发下拉刷新动画,效果与用户手动下拉刷新一致。它有 3 个回调函数:成功后回调函数、失败后回调函数、完成后回调函数,示例代码如下。

```
wx.startPullDownRefresh({
    success:function(res){
        //成功后回调函数
    },
    fail: function(res){
        //失败后回调函数
```

```
    },
    complete:{
        //完成后回调函数
    }
})
```

2. wx.stopPullDownRefresh 停止当前页面下拉刷新

微信小程序使用 wx.stopPullDownRefresh（Object object）来停止当前页面的下拉刷新。它有 3 个回调函数：成功后回调函数、失败后回调函数、完成后回调函数，示例代码如下。

```
Page({
    onPullDownRefresh() {
        wx.stopPullDownRefresh({
            success: function(res){
                //成功后回调函数
            },
            fail: function(res){
                //失败后回调函数
            },
            complete: function(res) {
                //完成后回调函数
            }
        })
    }
})
```

下面介绍下拉刷新的完整使用方法。

（1）在页面 demo.json 文件里配置下拉刷新属性，具体代码如下。

```
{
    "usingComponents": {},
    "backgroundTextStyle":"dark", //dark:显示刷新动画
    "enablePullDownRefresh":true, //允许下拉刷新
    "onReachBottomDistance":50//距离底部多少像素时触发上拉加载事件
}
```

（2）在页面 demo.js 文件里配置下拉刷新属性，具体代码如下。

```
Page({
    onLoad: function (options) {
        wx.startPullDownRefresh();//开始下拉刷新
    },

    /**
     * 页面相关事件处理函数--监听用户下拉动作
     */
    onPullDownRefresh: function () {
        wx.stopPullDownRefresh()//得到结果后关掉刷新动画
    },

    /**
     * 页面上拉触底事件的处理函数
     */
    onReachBottom: function () {
        //触发上拉相关操作
    },
})
```

效果如图 5.13 所示。

图 5.13 下拉刷新效果

5.6.2 wx.setBackgroundColor 动态设置窗口的背景色

微信小程序提供用 wx.setBackgroundColor（Object object）函数来动态设置窗口背景色的功能，可以整体设置背景色颜色、设置顶部背景色颜色或设置底部背景色颜色，Object 参数说明如表 5.9 所示。

表 5.9 wx.setBackgroundColor 的参数说明

属性	类型	是否必填	说明
backgroundColor	string	否	窗口的背景色，必须为十六进制颜色值
backgroundColorTop	string	否	顶部窗口的背景色，必须为十六进制颜色值，仅 iOS 支持
backgroundColorBottom	string	否	底部窗口的背景色，必须为十六进制颜色值，仅 iOS 支持
success	Function	否	接口调用成功的回调函数
fail	Function	否	接口调用失败的回调函数
complete	Function	否	接口调用结束的回调函数（调用成功、失败都会执行）

示例代码如下。

```
Page({
  onLoad: function (options) {
    wx.setBackgroundColor({
      backgroundColor: '#000000',   // 窗口的背景色为深黑色
    })

    wx.setBackgroundColor({
      //backgroundColorTop: '#999999',   // 顶部窗口的背景色为浅黑色
      //backgroundColorBottom: '#cccccc',   // 底部窗口的背景色为灰色
    })

    wx.startPullDownRefresh();//开始下拉刷新
  },

  /**
   * 页面相关事件处理函数--监听用户下拉动作
   */
  onPullDownRefresh: function () {
    wx.stopPullDownRefresh()//得到结果后关掉刷新动画
  },

  /**
   * 页面上拉触底事件的处理函数
   */
  onReachBottom: function () {
    //触发上拉相关操作
  },
})
```

效果如图 5.14 所示。

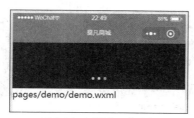

图 5.14 设置窗口背景色

5.6.3 wx.setBackgroundTextStyle 动态设置下拉背景字体

微信小程序提供用 wx.setBackgroundTextStyle（Object object）函数来动态设置下拉背景字体的功能，可以动态设置下拉背景字体、loading 图的样式，Object 参数说明如表 5.10 所示。

表 5.10 wx.setBackgroundTextStyle 的参数说明

属性	类型	是否必填	说明
textStyle	string	是	下拉背景字体、loading 图的样式，textStyle 可选值：dark/light（黑色/白色）
success	Function	否	接口调用成功的回调函数
fail	Function	否	接口调用失败的回调函数
complete	Function	否	接口调用结束的回调函数（调用成功、失败都会执行）

示例代码如下。

```
Page({
  onLoad: function (options) {
    wx.setBackgroundColor({
      backgroundColor: '#000000',  // 窗口的背景色为深黑色
    })

    wx.setBackgroundColor({
      //backgroundColorTop: '#999999',   // 顶部窗口的背景色为浅黑色
      //backgroundColorBottom: '#cccccc',  // 底部窗口的背景色为灰色
    })

    wx.setBackgroundTextStyle({
      textStyle: 'light'  // 下拉背景字体、loading 图的样式为 light(白色)
    })

    wx.startPullDownRefresh();//开始下拉刷新
  },

  /**
   * 页面相关事件处理函数——监听用户下拉动作
   */
  onPullDownRefresh: function () {
    wx.stopPullDownRefresh()//得到结果后关掉刷新动画
  },

  /**
   * 页面上拉触底事件的处理函数
   */
  onReachBottom: function () {
    //触发上拉相关操作
  },
})
```

效果如图 5.15 所示。

图 5.15 设置下拉背景字体

5.6.4 wx.loadFontFace 引入第三方字体

微信小程序提供引入第三方字体的 API 接口 wx.loadFontFace，可以动态加载网络字体。文件地址须为下载类型。iOS 仅支持 https 格式的文件地址。wx.loadFontFace（Object objec）的参数说明如表 5.11 所示。

表 5.11 wx. loadFontFace 的参数说明

属性	类型	是否必填	说明
family	string	是	定义的字体名称
source	string	是	字体资源的地址。建议格式为 TTF 和 WOFF，WOFF2 在低版本的 iOS 上会不兼容
desc	Object	否	可选的字体描述符
success	Function	否	接口调用成功的回调函数
fail	Function	否	接口调用失败的回调函数
complete	Function	否	接口调用结束的回调函数（调用成功、失败都会执行）

（1）可以在页面 JS 文件里引入第三方字体，示例代码如下。

```
Page({
  onLoad: function (options) {
    wx.loadFontFace({
      family: 'Bitstream Vera Serif Bold',
      source: 'url("https://sungd.github.io/Pacifico.ttf")',
      success: function(res){
        console.log(res); // {status: "loaded", cbID: 1}
      }
    })
  }
})
```

（2）在 WXSS 页面文件里，使用引入的字体，示例代码如下。

```
font-family: ' Bitstream Vera Serif Bold ';
```

（3）在引入第三方字体时，也可以新建一个字体文件 font.js 文件，以方便各个页面直接引用，示例代码如下。

```
//font.js
function loadFont(){
    wx.loadFontFace({
      family: 'Bitstream Vera Serif Bold',
      source: 'url("https://sungd.github.io/Pacifico.ttf")',
      success: function(res){
        console.log(res);
      }
    })
}
```

（4）在使用字体页面时，可以直接引入字体文件，示例代码如下。

```
const font = require('font.js')
Page({
```

```
onLoad: function (options) {
    font.loadFont(); //加载字体
  }
})
```

（5）在 WXSS 页面文件里，使用引入的字体，示例代码如下。

font-family: ' Bitstream Vera Serif Bold ';

注意：字体文件返回的内容类型 content-type 参考 font，格式不正确时会解析失败。字体链接的格式必须是 https（iOS 不支持 http）。字体链接必须是同源下的或开启了 cors 支持的，小程序的域名是 servicewechat.com。canvas 等原生组件不支持使用接口添加的字体。工具里提示的 Faild to load font 可以忽略。

5.6.5 wx.pageScrollTo 将页面滚动到目标位置

微信小程序 wx.pageScrollTo 接口 API 可以将页面滚动到目标位置，提供两种滚动方式：通过选择器的方式滚动和通过指定距离的方式滚动。通过这个 API 接口就可以实现长页面的回到顶部、回到底部功能，wx. pageScrollTo（Object objec）的参数说明如表 5.12 所示。

表 5.12 wx. pageScrollTo 的参数说明

属性	类型	是否必填	说明
scrollTop	number	否	滚动到页面的目标位置，单位为像素
duration	number	否	滚动动画的时长，单位为毫秒
selector	string	否	选择器
success	Function	否	接口调用成功的回调函数
fail	Function	否	接口调用失败的回调函数
complete	Function	否	接口调用结束的回调函数（调用成功、失败都会执行）

selector 选择器类似于 CSS 的选择器，但仅支持下列语法。

（1）ID 选择器：#the-id。

（2）class 选择器（可以连续指定多个）：.a-class.another-class。

（3）子元素选择器：.the-parent > .the-child。

（4）后代选择器：.the-ancestor .the-descendant。

（5）跨自定义组件的后代选择器：.the-ancestor >>> .the-descendant。

（6）多选择器的并集：#a-node，.some-other-nodes。

示例代码如下。

```
wx.pageScrollTo({
  scrollTop: 0,
  duration: 300
})
```

5.7 小结

本章讲解了微信小程序函数，包括生命周期函数、页面事件函数、页面路由管理、自定义函数、setData 设值函数；微信小程序网络请求，包括网络请求访问配置、wx.request 请求数据 API、wx.uploadFile 文件上传、wx.downloadFile 文件下载、WebSocket 会话 API；微信小程序模板的定义和使用；微信小程序的两种引用功能：import 引用和 include 引用；WXS 小程序脚本语言的使用；微信小程序提供的下拉刷新和监听事件的功能，使用 wx.setBackgroundColor 可以动态设置窗口的背景色，使用 wx.setBackgroundTextStyle 可以设置下拉背景文字，使用 wx.loadFontFace 可以引入第三方字体，使用 wx.pageScrollTo 可以将页面滚动到目标位置。

第6章
莫凡商城的注册、登录功能

注册、登录功能是非常通用的功能，几乎所有的网站、App、H5、小程序等都会用到注册、登录功能。莫凡商城的注册、登录等功能的设置会用到微信小程序的表单组件，涉及微信小程序界面交互、数据缓存 API 应用及微信小程序登录相关接口 API。在储备好这些基础知识之后，就可以实现莫凡商城注册功能、登录功能、修改密码功能、意见反馈功能和清除缓存功能的设置了。

慕课视频

微信小程序表单组件

6.1 微信小程序表单组件

微信小程序提供了丰富的表单组件，包括 button 按钮组件、checkbox 多选项目组件、radio 单选项目组件、form 表单组件、input 单行输入框组件、textarea 多行输入框组件、label 改进表单可用性组件、picker 从底部弹起的滚动选择器组件、switch 开关选择器组件、slider 滑动选择器组件。

6.1.1 button 按钮组件

button 按钮组件提供了 3 种类型的按钮：基本类型按钮、默认类型按钮及警告类型按钮，按钮的大小有默认和迷你两种，如图 6.1 所示。

图 6.1 按钮的类型和大小

button 按钮组件有很多属性，每个属性有不同的作用，如表 6.1 所示。

表 6.1 button 按钮组件的属性

属性	类型	默认值	说明
size	string	default	按钮的大小，有效值为 default、mini
type	string	default	按钮的样式类型，有效值：基本类型为 primary，默认类型为 default，警告类型为 warn

续表

属性	类型	默认值	说明
plain	boolean	false	按钮是否镂空,背景色透明
disabled	boolean	false	是否禁用
loading	boolean	false	名称前是否带 loading 图标
form-type	string	无	有效值为 submit、reset,用于<form/>组件,单击分别会触发 submit、reset 事件
open-type	string	否	微信开放能力,详见表后关于 open-type 合法值的介绍
hover-class	string	button-hover	指定按钮按下去的样式类。当 hover-class="none" 时,没有单击态效果
hover-stop-propagation	boolean	false	指定是否阻止本节点的祖先节点出现单击态
hover-start-time	number	20	按住后多久出现单击态,单位为毫秒
hover-stay-time	number	70	手指松开后单击态的保留时间,单位为毫秒
lang	string	en	指定返回用户信息的语言,zh_CN 为简体中文,zh_TW 为繁体中文,en 为英文
session-from	string		会话来源,open-type="contact"时有效
send-message-title	string	当前标题	会话内消息卡片标题,open-type="contact"时有效
send-message-path	string	当前分享路径	会话内消息卡片单击跳转小程序路径,open-type="contact"时有效
send-message-img	string	截图	会话内消息卡片图片,open-type="contact"时有效
app-parameter	string		打开 App 时,向 App 传递的参数,open-type=launchApp 时有效
show-message-card	boolean	false	是否显示会话内消息卡片,设置此参数为 true,用户进入客服会话会在右下角显示"可能要发送的小程序"提示,用户单击后可以快速发送小程序消息,open-type="contact"时有效
bindgetuserinfo	eventhandle		用户单击该按钮时,会返回获取到的用户信息,回调的 detail 数据与 wx.getUserInfo 返回的一致,open-type="getUserInfo"时有效
bindcontact	eventhandle		客服消息回调,open-type="contact"时有效
bindgetphonenumber	eventhandle		获取用户手机号回调,open-type=getPhonenumber 时有效
binderror	eventhandle		当使用开放能力时,发生错误的回调,open-type=launchApp 时有效
bindopensetting	eventhandle		在打开授权设置页后回调,open-type=openSetting 时有效
bindlaunchapp	eventhandle		打开App成功的回调,open-type=launchApp 时有效

open-type 的合法值详解如下。

(1)contact:打开客服会话,如果用户在会话中单击消息卡片后返回小程序,则可以从 bindcontact 回调中获得具体信息。

(2)share:触发用户转发,使用前建议先阅读使用指引。

(3)getPhonenumber:获取用户手机号,可以从 bindgetphonenumber 回调中获取到用户信息。

(4)getUserInfo:获取用户信息,可以从 bindgetuserinfo 回调中获取到用户信息。

(5)launchApp:打开 App,可以通过 app-parameter 属性设定向 App 传递参数。

(6)openSetting:打开授权设置页。

(7)feedback:打开"意见反馈"页面,用户可提交反馈内容并上传日志,开发者可以登录小程序管理后台后进入左侧菜单"客服反馈"页面获取到反馈内容。

从按钮属性中可以看出,按钮可以设置不同大小、不同类型、是否镂空、是否禁用、按钮名称前是否带 loading 图标等,针对 form 表单组件,按钮组件提供了提交表单和重置表单两个功能,具体代码如下。

```html
<button type="default" size="{{defaultSize}}" loading="{{loading}}" plain="{{plain}}"
        disabled="{{disabled}}" bindtap="default" style="margin:10px"> default </button>
<button type="primary" size="{{primarySize}}" loading="{{loading}}" plain="{{plain}}"
        disabled="{{disabled}}" bindtap="primary" style="margin:10px"> primary </button>
<button type="warn" size="{{warnSize}}" loading="{{loading}}" plain="{{plain}}"
        disabled="{{disabled}}" bindtap="warn" style="margin:10px"> warn </button>
<button bindtap="setDisabled" style="margin:10px">单击设置以上按钮 disabled 属性</button>
<button bindtap="setPlain" style="margin:10px">单击设置以上按钮 plain 属性</button>
<button bindtap="setLoading" style="margin:10px">单击设置以上按钮 loading 属性</button>
```

```javascript
var types = ['default', 'primary', 'warn']
var pageObject = {
  data: {
    defaultSize: 'default',
    primarySize: 'default',
    warnSize: 'default',
    disabled: false,
    plain: false,
    loading: false
  },
  setDisabled: function(e) {
    this.setData({
      disabled: !this.data.disabled
    })
  },
  setPlain: function(e) {
    this.setData({
      plain: !this.data.plain
    })
  },
  setLoading: function(e) {
    this.setData({
      loading: !this.data.loading
    })
  }
}

for (var i = 0; i < types.length; ++i) {
  (function(type) {
    pageObject[type] = function(e) {
      var key = type + 'Size'
      var changedData = {}
      changedData[key] =
        this.data[key] === 'default' ? 'mini' : 'default'
      this.setData(changedData)
    }
  })(types[i])
}

Page(pageObject)
```

界面效果如图 6.2 所示。

图 6.2　按钮效果

6.1.2　checkbox 多选项目组件

checkbox 多选项目组件，也叫多项选择器，同时也是我们常说的复选框，常用来进行多项选择，它的属性如表 6.2 所示。

表 6.2　checkbox 多选项目组件的属性

属性	类型	默认值	说明
value	string		<checkbox/>标识，选中时触发<checkbox-group/>的 change 事件，并携带<checkbox/>的 value
disabled	boolean	false	是否禁用
checked	boolean	false	当前是否选中，可用来设置默认选中
color	Color		checkbox 的颜色，同 CSS 的 color

checkbox-group 是用来容纳多个 checkbox 的多项选择器容器，它有一个绑定事件 bindchange，<checkbox-group/>中的选中项发生变化时会触发 change 事件，detail = {value: [选中的 checkbox 的 value 的数组]}。

下面演示一下 checkbox 多选项目组件的使用，以及获取选中的 value 值的方法。

（1）在 WXML 文件里使用 checkbox 进行界面布局，具体代码如下。

```
<view style="text-align:center;margin:10px;">
  <checkbox-group bindchange="checkboxChange">
    <checkbox value="篮球" />篮球
    <checkbox value="足球" checked="true" />足球
    <checkbox value="排球" />排球
    <checkbox value="橄榄球" disabled/>橄榄球
  </checkbox-group>
</view>
```

（2）在 JS 文件里，添加 checkboxChange 事件函数，获取复选框选中的值，将其打印出来，具体代码如下。

```
Page({
  checkboxChange:function(e){
      console.log(e.detail.value)
  }
})
```

界面效果如图 6.3 所示。

图 6.3 多选项目 value 值

从图 6.3 中可以看出，被禁用的多选项目是不能使用的，通过绑定 bindchange 事件，选项改变时会将多选项目的值以数组的形式存在 detail 里，通过 e.detail.value 就可以获取到多选项目的值。

多选项目的样式是可以重新定义的，可以不使用默认的效果，下面自定义多选项目的样式，添加 WXSS 样式代码。

```
/*checkbox 整体大小  */
checkbox {
    width: 200rpx;
    height: 80rpx;
}
/*checkbox 选项框大小  */
checkbox .wx-checkbox-input {
    width: 50rpx;
    height: 50rpx;
}
/*checkbox 选中后样式  */
checkbox .wx-checkbox-input.wx-checkbox-input-checked {
    background: #f50410;
}
/*checkbox 选中后图标样式  */
checkbox .wx-checkbox-input.wx-checkbox-input-checked::before {
    width: 28rpx;
    height: 28rpx;
    line-height: 28rpx;
    text-align: center;
    font-size: 22rpx;
    color: #fff;
    background: transparent;
    transform: translate(-50%, -50%) scale(1);
    -webkit-transform: translate(-50%, -50%) scale(1);
}
```

效果如图 6.4 所示。

图 6.4 修改多选项目样式

6.1.3 radio 单选项目组件

radio 单选项目组件，也叫单项选择器，是很常用的表单组件，每次只能选中一个选项，选项间是互斥关系，如用来选择性别，"男"或"女"选项只能选其一，它的属性如表 6.3 所示。

表 6.3 radio 单选项目组件的属性

属性	类型	默认值	说明
value	string		<radio/>标识。当该<radio/>被选中时，<radio-group/>的 change 事件会携带<radio/>的 value
disabled	boolean	false	是否禁用
checked	boolean	false	当前是否选中，可用来设置默认选中
color	Color		radio 的颜色，同 CSS 的 color

radio-group 是用来容纳多个 radio 单选项目组件的单项选择器容器，它有一个绑定事件 bindchange，<radio-group/> 中的选中项发生变化时会触发 bindchange 事件，event.detail = {value：选中项 radio 的 value}。

下面演示一下 radio 单选项目组件的使用。

（1）在 WXML 文件里使用 radio 单选项目进行界面布局，具体代码如下。

```
<view style="text-align:center;margin:10px;">
  <radio-group class="radio-group" bindchange="radioChange">
    <radio value="男" />男
    <radio value="女" checked/>女
    <radio value="未知" disabled/>未知
  </radio-group>
</view>
```

（2）在 JS 里，添加 radioChange 事件函数，获取单选项目选中的值，将其打印出来，具体代码如下。

```
Page({
  radioChange: function(e) {
    console.log('radio 发生 change 事件，携带 value 值为:', e.detail.value)
  }
})
```

界面效果如图 6.5 所示。

图 6.5 单选项目 value 值

从图 6.5 中可以看出，被禁用的单选项目是不能使用的，在 radio-group 上绑定 bindchange 事件，每次勾选时，只能使一个选项呈现为选中状态，同时会把相应的值存在 detail 里。

6.1.4 input 输入框组件

input 输入框组件是用来输入单行文本内容的，它的属性如表 6.4 所示。

表 6.4 input 输入框组件的属性

属性	类型	默认值	说明
value	string		输入框的初始内容
type	string	text	表示 input 的类型，有效值为 text（文本输入键盘）、number（数字输入键盘）、idcard（身份证输入键盘）、digit（带小数点的数字键盘）
password	boolean	false	是否是密码类型
placeholder	string		输入框为空时的占位符
placeholder-style	string		指定 placeholder 的样式
placeholder-class	string	input-placeholder	指定 placeholder 的样式类
disabled	boolean	false	是否禁用
maxlength	number	140	最大输入长度，设置为-1 时不限制最大长度
cursor-spacing	number	0	指定光标与键盘的距离，取 input 距离底部的距离和 cursor-spacing 指定的距离的最小值作为光标与键盘的距离
auto-focus	boolean	false	自动聚焦，拉起键盘（即将废弃，可直接使用 focus）
focus	boolean	false	获取焦点
confirm-type	string	done	设置键盘右下角按钮的文字，仅在 type='text'时生效，有效值为 send（发送）、search（搜索）、next（下一个）、go（前往）、done（完成）
confirm-hold	boolean	false	单击键盘右下角的按钮时是否保持键盘不收起
cursor	number		指定 focus 时的光标位置
selection-start	number	-1	光标起始位置，自动聚集时有效，需与 selection-end 搭配使用
selection-end	number	-1	光标结束位置，自动聚集时有效，需与 selection-start 搭配使用
adjust-position	boolean	true	键盘弹起时，是否自动上推页面
hold-keyboard	boolean	false	focus 时，单击页面时不收起键盘
bindinput	eventHandle		当用键盘输入时，触发 input 事件，event.detail = {value: value}，处理函数可以直接 return 一个字符串，将替换输入框的内容
bindfocus	eventHandle		输入框聚焦时触发，event.detail = {value: value}
bindblur	eventHandle		输入框失去焦点时触发，event.detail = {value: value}
bindconfirm	eventHandle		单击"完成"按钮时触发，event.detail = {value: value}
bindkeyboardheightchange	eventhandle		键盘高度发生变化时触发此事件，event.detail= {height: height, duration: duration}

从表 6.4 中可以看出以下几点。

（1）可以设置 input 输入框的类型，有 text 文本输入键盘、number 数字输入键盘、idcard 身份证输入键盘、digit 带小数点的数字键盘，可根据不同的场景使用不同的输入类型。

（2）可以设置输入框是否为密码类型，如果是密码类型，则会用点号代替具体值显示，这也是密码输入框的常用处理方式。

（3）通过 placeholder 来给输入框添加友好的提示信息，如"请输入手机号/用户名/邮箱"，用 placeholder-style 设置提示信息的样式，用 placeholder-class 设置提示信息的 class，然后再针对这个 class 添加样式。

（4）可以设置 input 输入框禁用和最大长度，并获取焦点。

（5）input 输入框有 3 个常用的事件：键盘输入时（bindinput）事件、输入框聚焦时（bindfocus）事件、失去焦点时（bindblur）事件。

示例代码如下。

（1）在 WXML 中利用 input 输入框进行布局，具体代码如下。

```
<view style="margin:10px">
  <view class="section">
    <input placeholder="这是一个可以自动聚焦的 input" auto-focus/>
  </view>
  <view class="section">
    <input placeholder="这个只有在单击按钮的时候才聚焦" focus="{{focus}}" />
    <view class="btn-area">
      <button bindtap="bindButtonTap">使得输入框获取焦点</button>
    </view>
  </view>
  <view class="section">
    <input maxlength="10" placeholder="最大输入长度 10" />
  </view>
  <view class="section">
    <view class="section__title">你输入的是:{{inputValue}}</view>
    <input bindinput="bindKeyInput" placeholder="输入同步到 view 中" />
  </view>
  <view class="section">
    <input bindinput="bindReplaceInput" placeholder="连续的两个 1 会变成 2" />
  </view>
  <view class="section">
    <input bindinput="bindHideKeyboard" placeholder="输入 123 自动收起键盘" />
  </view>
  <view class="section">
    <input password type="number" />
  </view>
  <view class="section">
    <input password type="text" />
  </view>
  <view class="section">
    <input type="digit" placeholder="带小数点的数字键盘" />
  </view>
  <view class="section">
    <input type="idcard" placeholder="身份证输入键盘" />
  </view>
  <view class="section">
    <input placeholder-style="color:red" placeholder="占位符字体是红色的" />
  </view>
</view>
```

（2）在 JS 文件中给 input 输入框添加相应的事件并提供数据，具体代码如下。

```
Page({
  data: {
    focus: false,
    inputValue: ''
  },
  bindButtonTap: function() {
    this.setData({
      focus: true
    })
  },
  bindKeyInput: function(e) {
    this.setData({
```

```
          inputValue: e.detail.value
        })
      },
      bindReplaceInput: function(e) {
        var value = e.detail.value
        var pos = e.detail.cursor
        if(pos != -1){
          //光标在中间
          var left = e.detail.value.slice(0, pos)
          //计算光标的位置
          pos = left.replace(/11/g, '2').length
        }

        //直接返回对象，可以对输入进行过滤处理，同时可以控制光标的位置
        return {
          value: value.replace(/11/g, '2'),
          cursor: pos
        }

        //或者直接返回字符串，光标在最后面
        //return value.replace(/11/g, '2'),
      },
      bindHideKeyboard: function(e) {
        if (e.detail.value === '123') {
          //收起键盘
          wx.hideKeyboard()
        }
      }
    })
```

界面效果如图 6.6 所示。

注意：input 组件是一个原生（native）组件，字体是系统字体，所以无法设置 font-family；在 input 聚焦期间，避免使用 CSS 动画；confirm-type 的最终表现与手机输入法本身的实现有关，可能不支持或不完全支持部分安卓系统输入法和第三方输入法；对于将 input 封装在自定义组件中，而 form 在自定义组件外的情况，form 将不能获得这个自定义组件中 input 的值，此时需要使用自定义组件的内置 behaviors wx://form-field；键盘高度发生变化，keyboardheightchange 事件可能会多次触发，开发者应该忽略掉相同的 height 值。

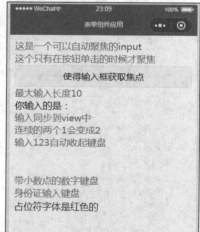

图 6.6　input 输入框

6.1.5　textarea 多行输入框组件

textarea 多行输入框组件是与 input 输入框组件对应的组件，它是用来输入多行文本内容的，它的属性如表 6.5 所示。

表 6.5　textarea 多行输入框组件的属性

属性	类型	默认值	说明
value	string		输入框的内容
placeholder	string		输入框为空时的占位符
placeholder-style	string		指定 placeholder 的样式，目前仅支持 color、font-size 和 font-weight
placeholder-class	string	input-placeholder	指定 placeholder 的样式类
disabled	boolean	false	是否禁用

续表

属性	类型	默认值	说明
maxlength	number	140	最大输入长度，设置为-1时不限制最大长度
auto-focus	boolean	false	自动聚焦，拉起键盘
focus	boolean	false	获取焦点
auto-height	boolean	false	是否自动增高，设置 auto-height 时，style.height 不生效
fixed	boolean	false	如果 textarea 在一个 position: fixed 的区域，需要显示指定属性 fixed 为 true
cursor-spacing	number	0	指定光标与键盘的距离。取 textarea 距离底部的距离和 cursor-spacing 指定的距离的最小值作为光标与键盘的距离
cursor	number	-1	指定 focus 时的光标位置
show-confirm-bar	boolean	true	是否显示键盘上方带有"完成"按钮的那一栏
selection-start	number	-1	光标起始位置，自动聚集时有效，需与 selection-end 搭配使用
selection-end	number	-1	光标结束位置，自动聚集时有效，需与 selection-start 搭配使用
adjust-position	boolean	true	键盘弹起时，是否自动上推页面
hold-keyboard	boolean	false	focus 时，单击页面时不收起键盘
bindkeyboardheightchange	eventhandle		键盘高度发生变化时触发此事件，event.detail = {height: height, duration: duration}
bindlinechange	eventHandle		输入框行数变化时调用，event.detail = {height: 0, heightRpx: 0, lineCount: 0}
bindinput	eventHandle		当用键盘输入时，触发 input 事件，event.detail = {value: value}，处理函数可以直接返回一个字符串，将替换输入框的内容
bindfocus	eventHandle		输入框聚焦时触发，event.detail = {value: value}
bindblur	eventHandle		输入框失去焦点时触发，event.detail = {value: value}
bindconfirm	eventHandle		单击"完成"按钮时触发，event.detail = {value: value}

从表6.5中可以看出以下几点。

（1）可以通过 placeholder 来给输入框添加友好的提示信息，用 placeholder-style 设置提示信息的样式，用 placeholder-class 设置提示信息的 class，然后再针对这个 class 添加样式。

（2）可以设置 textarea 输入框禁用和最大长度，并获取焦点，设置自动调整行高。

（3）textarea 输入框有4个常用的事件：键盘输入时（bindinput）事件、输入框聚焦时（bindfocus）事件、输入框失去焦点时（bindblur）事件、行数变化时（bindlinechange）事件。

示例代码如下：

```
<view class="section">
  <textarea bindblur="bindTextAreaBlur" auto-height placeholder="自动变高" />
</view>
<view class="section">
  <textarea placeholder="placeholder 颜色是红色的" placeholder-style="color:red;"  />
</view>
```

注意：textarea 的 blur 事件会晚于页面上的 tap 事件，如果需要在 button 的单击事件获取 textarea，可以使用 form 的 bindsubmit；不建议在多行文本上对用户的输入进行修改，所以 textarea 的 bindinput

处理函数并不会将返回值反映到 textarea 上；键盘高度发生变化，keyboardheightchange 事件可能会多次触发，开发者对于相同的 height 值应该忽略掉。

6.1.6 label 改进表单可用性组件

label 组件用来改进表单的可用性，目前可以用来改进的组件有<button/>、<checkbox/>、<radio/>和<switch/>。label 组件只有一个属性 for，用来绑定控件的 id。它的使用有两种方式：一种是没有定义 for 属性；另一种是定义了 for 属性。

1. label 组件没有定义 for 属性

label 组件没有定义 for 属性时，在 label 内包含<button/>、<checkbox/>、<radio/>、<switch/>这些组件，当单击 label 组件时，会触发 label 内包含的第一个控件，假如<button/>在第一个位置，就会触发<button/>对应的事件，假如<radio/>在第一位，就会触发 radio 对应的事件。

下面演示一下它的使用方法。

（1）在 WXML 文件里利用 label 组件布局，把第一个组件隐藏起来，具体代码如下。

```
<label>
    <button bindtap="clickBtn" hidden>我是 button 按钮</button>
    <view>我是 label 组件内的内容</view>
    <checkbox-group bindchange="checkboxChange">
        <checkbox value="中国" />中国
        <checkbox value="美国" />美国
    </checkbox-group>
    <radio-group bindchange="radioChange">
        <radio value="男"/>男
        <radio value="女"/>女
    </radio-group>
</label>
```

（2）在 JS 文件里添加 clickBtn、checkboxChange、radioChange 这 3 个事件函数，分别打印不同的信息，具体代码如下。

```
Page({
    clickBtn:function(){
        console.log("单击了按钮组件");
    },
    checkboxChange:function(){
        console.log("单击了多项选择器组件");
    },
    radioChange:function(){
        console.log("单击了单项选择器组件");
    }
})
```

（3）在 WXML 界面里可以看到<button/>按钮组件是隐藏起来的，但单击"我是 label 组件内的内容"，可以看到打印信息是按钮事件函数打印的信息，如图 6.7 所示。

图 6.7 没有定义 for 属性

从这里可以看出，label 组件内有多个组件时，会触发第一个组件。

2. label 组件定义了 for 属性

label 组件定义了 for 属性后，它会根据 for 属性的值找到和组件 id 一样的值，然后会触发这个组件的相应事件。

下面演示一下它的使用方法。

（1）在 WXML 文件里利用 label 组件布局，把第一个组件隐藏起来，给 label 定义 for 等于 man，让它找到 id 值等于 man 的组件，然后触发该组件的事件，具体代码如下。

```
<label for="man">
    <button id="btn" bindtap="clickBtn" hidden>我是 button 按钮</button>
    <view>我是 label 组件内的内容</view>
    <checkbox-group bindchange="checkboxChange" id="checkbox">
        <checkbox value="中国" />中国
        <checkbox value="美国" />美国
    </checkbox-group>
    <radio-group bindchange="radioChange" >
        <radio id="man" value="男"/>男
        <radio id="women" value="女"/>女
    </radio-group>
</label>
```

（2）在 JS 文件里添加 clickBtn、checkboxChange、radioChange 这 3 个事件函数，分别打印不同的信息，具体代码如下。

```
Page({
    clickBtn:function(){
        console.log("单击了按钮组件");
    },
    checkboxChange:function(){
        console.log("单击了多项选择器组件");
    },
    radioChange:function(){
        console.log("单击了单项选择器组件");
    }
})
```

（3）在 WXML 页面里可以看到<button/>按钮组件是隐藏起来的，但是单击"我是 label 组件内的内容"，可以看到 id 值等于 man 的单项选择器组件呈现为选中状态，同时触发事件，打印信息，如图 6.8 所示。

图 6.8　定义了 for 属性

综上所述，如果 label 定义了 for 属性，它会根据 for 属性的值找到和组件 id 一样的值，然后触发相应事件；如果 label 没有定义 for 属性，它会找到 label 组件内的第一个组件，然后触发相应事件。

6.1.7　picker 滚动选择器组件

picker 滚动选择器组件主要支持 5 种滚动选择器：普通选择器、时间选择器、日期选择器、多列选择器和省市区选择器，默认是普通选择器。另外，还有一种picker-view嵌入页面滚动选择器，支持将滚动选择器嵌入页面中。

以上5种滚动选择器是通过 mode 来区分的：普通选择器 mode=selector、时间选择器 mode=time、日期选择器 mode=date、多列选择器 mode=multiSelector、省市区选择器 mode=region，每种类型选择器的属性不同。

1. 普通选择器：mode = selector

普通选择器的属性如表6.6所示。

表6.6　普通选择器的属性

属性	类型	默认值	说明
range	array/object array	[]	mode 为 selector 或 multiSelector 时，范围（range）有效
range-key	string		当范围（range）是一个对象数组（Object Array）时，通过 range-key 来指定 Object 中 key 的值作为选择器的显示内容
value	number	0	表示选择了 range 中的第几个（下标从0开始）
bindchange	eventhandle		value 改变时触发 change 事件，event.detail = {value}

示例代码如下。

```
<view class="section">
  <view class="section__title">地区选择器</view>
  <picker bindchange="bindPickerChange" value="{{index}}" range="{{array}}">
    <view class="picker">
      当前选择:{{array[index]}}
    </view>
  </picker>
</view>
```

```
Page({
  data: {
    array: ['美国', '中国', '巴西', '日本'],
    objectArray: [
      {
        id: 0,
        name: '美国'
      },
      {
        id: 1,
        name: '中国'
      },
      {
        id: 2,
        name: '巴西'
      },
      {
        id: 3,
        name: '日本'
      }
    ],
    index: 0
  },
  bindPickerChange: function(e) {
    console.log('picker发送选择改变，携带值为', e.detail.value)
    this.setData({
      index: e.detail.value
```

      })
    }
})
界面效果如图 6.9 所示。

图 6.9　普通选择器

2. 时间选择器：mode = time

时间选择器的属性如表 6.7 所示。

表 6.7　时间选择器的属性

属性	类型	说明
value	string	表示选中的时间，格式为"hh: mm"
start	string	表示有效时间范围的开始，字符串格式为"hh: mm"
end	string	表示有效时间范围的结束，字符串格式为"hh: mm"
bindchange	eventHandle	value 改变时触发 change 事件，event.detail = {value}

示例代码如下。

```
<view class="section">
  <view class="section__title">时间选择器</view>
  <picker mode="time" value="{{time}}" start="09:01" end="21:01" bindchange="bindTimeChange">
    <view class="picker">
      当前选择: {{time}}
    </view>
  </picker>
</view>

Page({
  data: {
    time: '12:01'
  },
  bindTimeChange: function(e) {
    this.setData({
```

```
          time: e.detail.value
      })
   }
})
```
界面效果如图 6.10 所示。

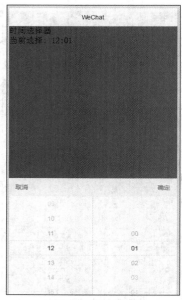

图 6.10　时间选择器

3. 日期选择器：mode = date

日期选择器的属性如表 6.8 所示。

表 6.8　日期选择器的属性

属性	类型	默认值	说明
value	string	0	表示选中的日期，格式为"YYYY-MM-DD"
start	string		表示有效日期范围的开始，字符串格式为"YYYY-MM-DD"
End	string		表示有效日期范围的结束，字符串格式为"YYYY-MM-DD"
Fields	string	Day	有效值为 year、month、day，表示选择器的粒度
Bindchange	eventhandle		value 改变时触发 change 事件，event.detail = {value}

示例代码如下。

```
<view class="section">
   <view class="section__title">日期选择器</view>
   <picker mode="date" value="{{date}}" start="2018-09-01" end="2021-09-01" bindchange="bindDateChange">
      <view class="picker">
         当前选择: {{date}}
      </view>
   </picker>
</view>
```

```
Page({
   data: {
      date: '2020-09-01'
   },
```

```
bindDateChange: function(e) {
  this.setData({
    date: e.detail.value
  })
}
})
```

界面效果如图 6.11 所示。

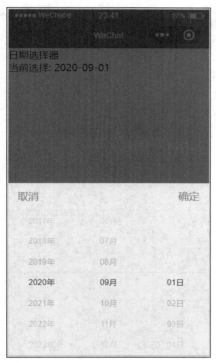

图 6.11　日期选择器

4. 多列选择器：mode = multiSelector

多列选择器的属性如表 6.9 所示。

表 6.9　多列选择器的属性

属性	类型	默认值	说明
range	array/object array	[ ]	mode 为 selector 或 multiSelector 时，范围（range）有效
range-key	string		当范围（range）是一个对象数组（Object Array）时，通过 range-key 来指定 Object 中 key 的值作为选择器的显示内容
value	number	0	表示选择了 range 中的第几个（下标从 0 开始）
bindchange	eventhandle		value 改变时触发 change 事件，event.detail = {value}
bindcolumnchange	eventhandle		列改变时触发

示例代码如下。

```
<view class="section">
  <view class="section__title">多列选择器</view>
  <picker mode="multiSelector" bindchange="bindMultiPickerChange" bindcolumnchange="bindMultiPickerColumnChange" value="{{multiIndex}}" range="{{multiArray}}">
```

```
        <view class="picker">
            当前选择:{{multiArray[0][multiIndex[0]]}}, {{multiArray[1][multiIndex[1]]}}, {{multiArray[2][multiIndex[2]]}}
        </view>
    </picker>
</view>
```

```
Page({
    data: {
        multiArray: [['无脊椎动物', '脊椎动物'], ['扁形动物', '线形动物', '环节动物', '软体动物', '节肢动物'], ['猪肉绦虫', '吸血虫']],
        objectMultiArray: [
            [
                {
                    id: 0,
                    name: '无脊柱动物'
                },
                {
                    id: 1,
                    name: '脊柱动物'
                }
            ], [
                {
                    id: 0,
                    name: '扁形动物'
                },
                {
                    id: 1,
                    name: '线形动物'
                },
                {
                    id: 2,
                    name: '环节动物'
                },
                {
                    id: 3,
                    name: '软体动物'
                },
                {
                    id: 3,
                    name: '节肢动物'
                }
            ], [
                {
                    id: 0,
                    name: '猪肉绦虫'
                },
                {
                    id: 1,
                    name: '吸血虫'
                }
            ]
        ],
        multiIndex: [0, 0, 0]
    },
    bindMultiPickerChange: function (e) {
        console.log('picker 发送选择改变，携带值为', e.detail.value)
        this.setData({
            multiIndex: e.detail.value
        })
    },
    bindMultiPickerColumnChange: function (e) {
```

```javascript
        console.log('修改的列为', e.detail.column, '，值为', e.detail.value);
        var data = {
          multiArray: this.data.multiArray,
          multiIndex: this.data.multiIndex
        };
        data.multiIndex[e.detail.column] = e.detail.value;
        switch (e.detail.column) {
          case 0:
            switch (data.multiIndex[0]) {
              case 0:
                data.multiArray[1] = ['扁形动物', '线形动物', '环节动物', '软体动物', '节肢动物'];
                data.multiArray[2] = ['猪肉绦虫', '吸血虫'];
                break;
              case 1:
                data.multiArray[1] = ['鱼', '两栖动物', '爬行动物'];
                data.multiArray[2] = ['鲫鱼', '带鱼'];
                break;
            }
            data.multiIndex[1] = 0;
            data.multiIndex[2] = 0;
            break;
          case 1:
            switch (data.multiIndex[0]) {
              case 0:
                switch (data.multiIndex[1]) {
                  case 0:
                    data.multiArray[2] = ['猪肉绦虫', '吸血虫'];
                    break;
                  case 1:
                    data.multiArray[2] = ['蛔虫'];
                    break;
                  case 2:
                    data.multiArray[2] = ['蚂蚁', '蚂蟥'];
                    break;
                  case 3:
                    data.multiArray[2] = ['河蚌', '蜗牛', '蛞蝓'];
                    break;
                  case 4:
                    data.multiArray[2] = ['昆虫', '甲壳动物', '蛛形动物', '多足动物'];
                    break;
                }
                break;
              case 1:
                switch (data.multiIndex[1]) {
                  case 0:
                    data.multiArray[2] = ['鲫鱼', '带鱼'];
                    break;
                  case 1:
                    data.multiArray[2] = ['青蛙', '娃娃鱼'];
                    break;
                  case 2:
                    data.multiArray[2] = ['蜥蜴', '龟', '壁虎'];
                    break;
                }
                break;
            }
            data.multiIndex[2] = 0;
            break;
        }
        console.log(data.multiIndex);
        this.setData(data);
      }
    })
```

界面效果如图 6.12 所示。

图 6.12　多列选择器

5. 省市区选择器：mode = region

省市区选择器的属性如表 6.10 所示。

表 6.10　省市区选择器的属性

属性	类型	默认值	说明
value	string	[ ]	表示选中的省市区，默认选中每一列的第一个值
custom-item	string		可为每一列的顶部添加一个自定义的项
bindchange	eventhandle		value 改变时触发 change 事件，event.detail = {value, code, postcode}，其中字段 code 是统计用区划代码，postcode 是邮政编码

示例代码如下。

```
<view class="section">
  <view class="section__title">省市区选择器</view>
  <picker mode="region" bindchange="bindRegionChange" value="{{region}}" custom-item="{{customItem}}">
    <view class="picker">
      当前选择:{{region[0]}}, {{region[1]}}, {{region[2]}}
    </view>
  </picker>
</view>
```

```
Page({
  data: {
    region: ['广东省', '广州市', '海珠区'],
    customItem: '全部'
  },
  bindRegionChange: function (e) {
    console.log('picker 发送选择改变，携带值为', e.detail.value)
    this.setData({
      region: e.detail.value
```

      })
    }
  })
```

界面效果如图 6.13 所示。

**6．picker-view 嵌入页面滚动选择器**

除了以上 5 种滚动选择器之外，还有一种嵌入页面的滚动选择器，可以使用 picker-view 组件在页面里进行布局，如图 6.14 所示。

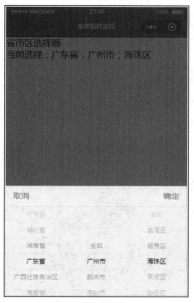

图 6.13　省市区选择器

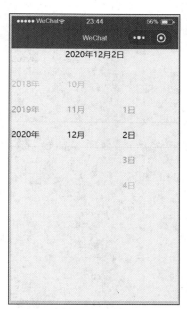

图 6.14　嵌入页面滚动选择器

picker-view 中只能使用<picker-view-column/>组件，picker-view 有 3 个属性，如表 6.11 所示。

表 6.11　嵌入页面滚动选择器的属性

| 属性 | 类型 | 说明 |
| --- | --- | --- |
| value | Array.&lt;number&gt; | 数组中的数字依次表示 picker-view 内的 picker-view-column 选择的第几项（下标从 0 开始），数字大于 picker-view-column 可选项长度时，选择最后一项 |
| indicator-style | String | 设置选择器中间选中框的样式 |
| indicator-class | String | 设置选择器中间选中框的类名 |
| mask-style | String | 设置蒙层的样式 |
| mask-class | String | 设置蒙层的类名 |
| bindchange | eventHandle | 滚动选择时触发 change 事件，event.detail = {value}；value 为数组，表示 picker-view 内的 picker-view-column 当前选择的是第几项（下标从 0 开始） |
| bindpickstart | Eventhandle | 当滚动选择开始时触发事件 |
| bindpickend | Eventhandle | 当滚动选择结束时触发事件 |

示例代码如下。

```
<view>
  <view style="text-align:center">{{year}}年{{month}}月{{day}}日</view>
```
```

```
<picker-view indicator-style="height: 50px;" style="width: 100%; height: 300px;" value="{{value}}" bindchange="bindChange">
    <picker-view-column>
      <view wx:for="{{years}}" style="line-height: 50px">{{item}}年</view>
    </picker-view-column>
    <picker-view-column>
      <view wx:for="{{months}}" style="line-height: 50px">{{item}}月</view>
    </picker-view-column>
    <picker-view-column>
      <view wx:for="{{days}}" style="line-height: 50px">{{item}}日</view>
    </picker-view-column>
  </picker-view>
</view>
```

```
const date = new Date()
const years = [ ]
const months = [ ]
const days = [ ]

for (let i = 1990; i <= date.getFullYear(); i++) {
   years.push(i)
}

for (let i = 1 ; i <= 12; i++) {
   months.push(i)
}

for (let i = 1 ; i <= 31; i++) {
   days.push(i)
}

Page({
  data: {
    years: years,
    year: date.getFullYear(),
    months: months,
    month: 2,
    days: days,
    day: 2,
    year: date.getFullYear(),
    value: [9999, 1, 1],
  },
  bindChange: function(e) {
    const val = e.detail.value
    this.setData({
      year: this.data.years[val[0]],
      month: this.data.months[val[1]],
      day: this.data.days[val[2]]
    })
  }
})
```

6.1.8 slider 滑动选择器组件

slider 滑动选择器组件经常在控制声音的大小、屏幕的亮度等场景中使用，它可以设置滑动步长，显示当前值并设置最小值和最大值，如图 6.15 所示。

图 6.15 滑动选择器

slider 滑动选择器组件的属性如表 6.12 所示。

表 6.12 slider 滑动选择器的属性

属性	类型	默认值	说明
min	number	0	最小值
max	number	100	最大值
step	number	1	步长，取值必须大于 0，并且可被（max-min）整除
disabled	boolean	false	是否禁用
value	number	0	当前取值
color	Color	#e9e9e9	背景条的颜色（请使用 backgroundColor）
selected-color	Color	#1aad19	已选择的颜色（请使用 activeColor）
activeColor	color	#1aad19	已选择的颜色
backgroundColor	color	#e9e9e9	背景条的颜色
block-size	number	28	滑块的大小，取值范围为 12~28
block-color	color	#ffffff	滑块的颜色
show-value	boolean	false	是否显示当前 value
bindchange	eventhandle		完成一次拖动后触发的事件，event.detail = {value: value}
bindchanging	eventhandle		拖动过程中触发的事件，event.detail = {value}

示例代码如下。

```
<view class="section section_gap">
  <text class="section__title">设置 step</text>
  <view class="body-view">
    <slider bindchange="sliderchange" step="5"/>
  </view>
</view>

<view class="section section_gap">
  <text class="section__title">显示当前 value</text>
  <view class="body-view">
    <slider bindchange="sliderchange" show-value/>
  </view>
</view>

<view class="section section_gap">
  <text class="section__title">设置最小V最大值</text>
  <view class="body-view">
    <slider bindchange="sliderchange" min="50" max="200" show-value/>
  </view>
</view>

<view class="section section_gap">
```

```
    <text class="section__title">设置颜色</text>
    <view class="body-view">
      <slider bindchange="sliderchange"  color="black" selected-color="red"/>
    </view>
</view>

<view class="section section_gap">
    <text class="section__title">禁用</text>
    <view class="body-view">
      <slider bindchange="sliderchange" disabled show-value/>
    </view>
</view>
```

界面效果如图 6.16 所示。

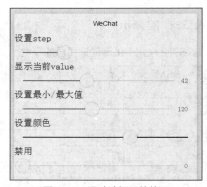

图 6.16　滑动选择器的使用

6.1.9　switch 开关选择器组件

switch 开关选择器组件有两个状态：开、关，在很多场景中都会用到这个功能，如微信设置里的新消息提醒界面，就通过开关来设置是否接收消息、是否显示消息、是否有声音、是否震动等，如图 6.17 所示。

图 6.17　微信新消息提醒设置

switch 开关选择器的属性包括是否选中、开关类型、颜色及绑定事件，如表 6.13 所示。

表 6.13 switch 开关选择器的属性

属性	类型	默认值	说明
checked	boolean	false	是否选中
disabled	boolean	false	是否禁用
type	string	switch	样式，有效值为 switch 和 checkbox
bindchange	eventhandle		checked 改变时触发 change 事件，event.detail={ value}
color	string	#04BE02	switch 的颜色，同 CSS 中的 color

示例代码如下。

```
<view style="background-color:#cccccc;height:600px;">
  <view style="padding-top:10px;"></view>
  <view style="display:flex;flex-direction:row;background-color:#ffffff;height:50px;line-height:50px;">
    <view style="font-weight:bold;">接收新消息通知</view>
    <view style="position:absolute;right:10px;">
      <switch type="switch" checked/>
    </view>
  </view>
  <view style="height:1px;background-color:#f2f2f2;opacity:0.2"></view>
  <view style="display:flex;flex-direction:row;background-color:#ffffff;height:50px;line-height:50px;">
    <view style="font-weight:bold;">通知显示消息详情</view>
    <view style="position:absolute;right:10px;">
      <switch type="switch"/>
    </view>
  </view>
  <view style="height:1px;background-color:#f2f2f2;opacity:0.2"></view>

  <view style="margin-top:20px;"></view>
  <view style="height:1px;background-color:#f2f2f2;opacity:0.2"></view>
  <view style="display:flex;flex-direction:row;background-color:#ffffff;height:50px;line-height:50px;">
    <view style="font-weight:bold;">声音</view>
    <view style="position:absolute;right:10px;">
      <switch type="checkbox" checked/>
    </view>
  </view>
  <view style="height:1px;background-color:#f2f2f2;opacity:0.2"></view>
  <view style="height:1px;background-color:#f2f2f2;opacity:0.2"></view>
  <view style="display:flex;flex-direction:row;background-color:#ffffff;height:50px;line-height:50px;">
    <view style="font-weight:bold;">震动</view>
    <view style="position:absolute;right:10px;">
      <switch type="checkbox"/>
    </view>
  </view>
  <view style="height:1px;background-color:#f2f2f2;opacity:0.2"></view>
</view>
```

界面效果如图 6.18 所示。

图 6.18 开关选择器的应用

6.1.10 form 表单组件

form 表单组件是用来将表单中的值提交给 JS 文件进行处理的，它可以提交 <switch/> <input/> <checkbox/> <slider/> <radio/> <picker/> 这些组件的值，提交表单的时候，会借助于 button 组件的 formType 为 submit 的属性，将表单组件中的 value 值进行提交，需要在表单组件中加上 name 来作为 key。form 表单组件的属性如表 6.14 所示。

表 6.14 form 表单的属性

属性	类型	默认值	说明
report-submit	boolean	false	是否返回 formId 用于发送模板消息
report-submit-timeout	number	0	等待一段时间（毫秒数）以确认 formId 是否生效。如果未指定这个参数，formId 有很小的概率是无效的（如遇到网络失败的情况）。指定这个参数将可以检测 formId 是否有效，以这个参数的时间作为这项检测的超时时间。如果失败，将返回 requestFormId: fail 开头的 formId
bindsubmit	eventhandle		携带 form 中的数据触发 submit 事件，event.detail = {value: {'name': 'value'}, formId: ''}
bindreset	eventhandle		表单重置时会触发 reset 事件

示例代码如下。

```
<form bindsubmit="formSubmit" bindreset="formReset">
  <view style="margin:10px;">
    <view style="font-weight:bold;">switch 开关选择器</view>
    <switch name="switch"/>
  </view>
  <view style="margin:10px;">
    <view style="font-weight:bold;">slider 滑动选择器</view>
    <slider name="slider" show-value ></slider>
  </view>
  <view style="margin:10px;">
    <view style="font-weight:bold;">input 单行输入框</view>
    <input name="input" placeholder="please input here" />
  </view>
  <view style="margin:10px;">
    <view style="font-weight:bold;">radio 单项选择器</view>
    <radio-group name="radio-group">
      <label><radio value="radio1"/>radio1</label>
      <label><radio value="radio2"/>radio2</label>
    </radio-group>
  </view>
  <view style="margin:10px;">
    <view style="font-weight:bold;">checkbox 多项选择器</view>
    <checkbox-group name="checkbox">
      <label><checkbox value="checkbox1"/>checkbox1</label>
      <label><checkbox value="checkbox2"/>checkbox2</label>
    </checkbox-group>
  </view>
  <view class="btn-area">
    <button formType="submit" type="primary">Submit</button>
    <button formType="reset">Reset</button>
  </view>
</form>
```

```
Page({
  formSubmit: function(e) {
```

```
      console.log('form 发生了 submit 事件，携带数据为:', e.detail.value)
    },
    formReset: function() {
      console.log('form 发生了 reset 事件')
    }
})
```

界面效果如图 6.19 和图 6.20 所示。

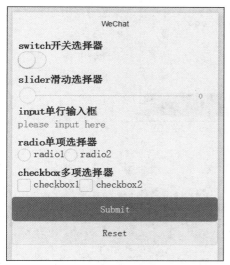

图 6.19　未填写表单

图 6.20　填写表单

单击"Reset"按钮可以重置表单，单击"Submit"按钮组件，就可以把表单数据提交到 JS 文件里进行处理，图 6.20 提交的数据如图 6.21 所示。

图 6.21　表单提交的数据

6.1.11　项目实战：任务 2——实现注册功能

1. 任务目标

综合应用容器组件和表单组件来实现注册功能。莫凡商城的注册页面包含用户名、手机号、密码、确认密码、昵称 5 个字段，如图 6.22 所示。

2. 任务实施

下面我们一起来实现莫凡商城的注册功能。
（1）在 app.json 文件里添加注册页面路径"pages/register/register"。
（2）在 register.wxml 页面文件里进行表单布局设计，包括 5 个字段，具体代码如下。

图6.22 注册页面

```
<form bindsubmit="formSubmit" bindreset="formReset">
  <view class="content">
    <view class="loginTitle">
      创建账号
    </view>
    <view class="hr"></view>
    <view class="accountType">
      <view class="account">
        <view class="ac">用户名</view>
        <view class="ipt"><input name="loginName" type="text" placeholder="请输入用户名" class="placeholder-style"/></view>
      </view>
      <view class="hr"></view>
      <view class="account">
        <view class="ac">手机号</view>
        <view class="ipt"><input name="mobile" type="text" placeholder="请输入手机号" class="placeholder- style"/></view>
      </view>
      <view class="hr"></view>
      <view class="account">
        <view class="ac">密码</view>
        <view class="ipt"><input name="loginPassword" type="text" password placeholder="请输入密码" class="placeholder-style"/></view>
      </view>
      <view class="hr"></view>
      <view class="account">
        <view class="ac">确认密码</view>
        <view class="ipt"><input name="confirmPassword" type="text" password placeholder="请确认密码" class="placeholder-style"/></view>
      </view>
      <view class="hr"></view>
      <view class="account">
        <view class="ac">昵称</view>
        <view class="ipt"><input name="nickName" type="text" placeholder="请输入昵称" class="placeholder-style"/></view>
      </view>
      <view class="hr"></view>
      <view class="login">
        <button form-type="submit">注册</button>
        <view class="tip">{{tip}}</view>
      </view>
```

```
        </view>
    </view>
</form>
```

（3）在 register.wxss 页面文件里进行表单样式设计，具体代码如下。

```css
.content{
    height: 600px;
}
.loginTitle{
    margin: 10px;
    text-align: center;
}
.select{
    font-size:12px;
    color: red;
    width: 50%;
    text-align: center;
    height: 45px;
    line-height: 45px;
    border-bottom:5rpx solid red;
}
.default{
    font-size:12px;
    margin: 0 auto;
    padding: 15px;
}
.hr{
    border: 1px solid #cccccc;
    opacity: 0.2;
}
.account{
    display: flex;
    flex-direction: row;
    align-items: center;
}
.ac{
    padding:15px;
    font-size:14px;
    font-weight: bold;
    color: #666666;
    width: 60px;
    text-align: center;
}
.ipt input{
    text-align: left;
    width: 200px;
    color: #000000;
}
.placeholder-style{
    font-size: 14px;
    color: #cccccc;
}
.login{
    margin: 0 auto;
    text-align: center;
    padding-top:10px;
}
.login button{
  width: 96%;
  color: #ffffff;
  background: #009966;
```

```css
}
.tip{
  margin-top:10px;
  font-size: 12px;
  color: #D53E37;
}
```

（4）在 register.js 页面文件里添加表单字段验证函数和表单提交函数，并发起网络请求，将表单数据提交到后台服务器以实现注册功能，具体代码如下。

```javascript
var app = getApp();
var host = app.globalData.host;
Page({
  data: {
    tip: ''//提示信息
  },
  formSubmit: function (e) {//提交表单
    var that = this;
    var loginName = e.detail.value.loginName;
    var mobile = e.detail.value.mobile;
    var loginPassword = e.detail.value.loginPassword;
    var confirmPassword = e.detail.value.confirmPassword;
    var nickName = e.detail.value.nickName;
    //验证表单输入
    var ret = that.checkUser(loginName, mobile, loginPassword, confirmPassword, nickName);
    if(ret){
      wx.request({
        url: host + '/api/user/register',
        method: 'GET',
        data: { 'loginName': loginName, 'mobile': mobile, 'loginPassword': loginPassword, 'confirmPassword': confirmPassword, 'nickName': nickName },
        header: {
          'Content-Type': 'application/json'
        },
        success: function (res) {
          var code = res.data.code;
          var msg = res.data.data;
          if (code == '0000') {
            wx.redirectTo({
              url: '../login/login'
            })
          } else {
            that.setData({ tip: msg });
            return false
          }
        }
      })
    }
  },
  checkUser: function(loginName, mobile, loginPassword, confirmPassword, nickName){//验证表单
    var that = this;
    if (loginName == "") {
      that.setData({ tip: '用户名不能为空！' });
      return false
    }

    if (mobile == '') {
      that.setData({ tip: '手机号不能为空！' });
      return false
    }

    var myreg = /^[1][3, 4, 5, 7, 8][0-9]{9}$/;
```

```
      if (!myreg.test(mobile)) {
        that.setData({ tip: '手机号不合法！' });
        return false;
      }

      if (loginPassword == '') {
        that.setData({ tip: '密码不能为空！' });
        return false
      }

      if (confirmPassword == '') {
        that.setData({ tip: '确认密码不能为空！' });
        return false
      }

      if (loginPassword != confirmPassword) {
        that.setData({ tip: '两次密码输入不一致！' });
        return false
      }

      if (nickName == '') {
        that.setData({ tip: '昵称不能为空！' });
        return false
      }
      that.setData({ tip: '' });
      return true
    }
  })
```

莫凡商城注册功能效果如图 6.23 所示。

图 6.23 注册功能

综上，我们应用到了 view 容器组件、form 表单组件、input 输入框组件及表单验证、表单提交、发起网络请求等基础知识，实现了莫凡商城表单注册功能。

6.2 微信小程序界面交互 API

微信小程序界面交互 API 包括 wx.showToast 显示消息提示框 API、wx.hideToast 隐藏消息提示框 API、wx.showModal 显示模态对话框 API、wx.showLoading 显示 loading 提示框

API、wx.hideLoading 隐藏 loading 提示框 API、wx.showActionSheet 显示操作菜单 API。通过使用这些界面交互 API，可以给用户创造友好的使用体验，提供提示信息。

6.2.1 wx.showToast/wx.hideToast 显示/隐藏消息提示框 API

wx.showToast 用来显示消息提示框，wx.hideToast 用来隐藏消息提示框，wx.showToast（Object object）的参数说明如表 6.15 所示。

表6.15 wx.showToast 的参数说明

属性	类型	默认值	说明
title	string	false	提示的内容
icon	string	'success'	图标，包括 success 显示成功图标，此时 title 文本最多显示 7 个汉字长度；loading 显示加载图标，此时 title 文本最多显示 7 个汉字长度；none 不显示图标，此时 title 文本最多可显示两行
image	string		自定义图标的本地路径，image 的优先级高于 icon
duration	number	1500	提示的延迟时间
mask	boolean	false	是否显示透明蒙层，防止触摸穿透
success	function		接口调用成功的回调函数
fail	function		接口调用失败的回调函数
complete	function		接口调用结束的回调函数（调用成功、失败都会执行）

示例代码如下。

```
Page({
  onLoad:function(){
    //显示成功提示信息
    wx.showToast({
      title: '成功',
      icon: 'success',
      duration: 2000
    })

    //显示加载中提示信息
    wx.showToast({
      title: '加载中',
      icon: 'loading',
      duration: 2000
    })

    //不显示 icon 提示信息
    wx.showToast({
      title: '不显示 icon',
      icon: 'none',
      duration: 2000
    })

    //图片图标提示信息
    wx.showToast({
      title: '图片图标',
      icon: 'none',
      image:'../images/icon/payyes.jpg',
      duration: 2000
    })

    //隐藏提示信息
    wx.hideToast();
  }
})
```

效果如图 6.24～图 6.27 所示。

图 6.24　success

图 6.25　loading

图 6.26　none

图 6.27　图片图标

6.2.2　wx.showModal 显示模态对话框 API

wx.showModal API 用来显示模态对话框，wx.showModal(Object object)的参数说明如表 6.16 所示。

表 6.16　wx.showModal 的参数说明

属性	类型	默认值	说明
title	string	false	提示的标题
content	string		提示的内容
showCancel	boolean	true	是否显示"取消"按钮
cancelText	string	'取消'	"取消"按钮的文字，最多 4 个字符
cancelColor	string	#000000	"取消"按钮的文字颜色，必须是 16 进制格式的颜色字符串
confirmText	string	'确定'	"确认"按钮的文字，最多 4 个字符
confirmColor	string	#576B95	"确认"按钮的文字颜色，必须是 16 进制格式的颜色字符串
success	function		接口调用成功的回调函数
fail	function		接口调用失败的回调函数
complete	function		接口调用结束的回调函数（调用成功、失败都会执行）

示例代码如下。

```
Page({
  onLoad:function(){
    wx.showModal({
      title: '温馨提示',
      content: '这是一个模态对话框',
      success(res) {
        if (res.confirm) {
          console.log('用户单击确定')
        } else if (res.cancel) {
          console.log('用户单击取消')
        }
      }
    })
  }
})
```

效果如图 6.28 所示。

图 6.28　模态对话框

6.2.3 wx.showLoading/wx.hideLoading 显示/隐藏 loading 提示框 API

wx.showLoading API 用来显示 loading 提示框，wx.hideLoading API 用来隐藏 loading 提示框，wx.showLoading（Object object）的参数说明如表 6.17 所示。

表 6.17 wx. showLoading 的参数说明

属性	类型	默认值	说明
title	string	false	提示的内容
mask	boolean	false	是否显示透明蒙层，防止触摸穿透
success	function		接口调用成功的回调函数
fail	function		接口调用失败的回调函数
complete	function		接口调用结束的回调函数（调用成功、失败都会执行）

示例代码如下。

```
Page({
  onLoad:function(){
    //显示 loading 提示框
    wx.showLoading({
      title: '加载中',
    })

    //隐藏 loading 提示框
    wx.hideLoading();
  }
})
```

效果如图 6.29 所示。

图 6.29 loading 提示框

6.2.4 wx.showActionSheet 显示操作菜单 API

wx.showActionSheet API 用来显示操作菜单，菜单最多显示 6 个，wx.showActionSheet（Object object）的参数说明如表 6.18 所示。

表 6.18 wx. showActionSheet 的参数说明

属性	类型	默认值	说明
itemList	Array.<string>		按钮的文字数组，数组长度最大为 6
itemColor	string	#000000	按钮的文字颜色
success	function		接口调用成功的回调函数
fail	function		接口调用失败的回调函数
complete	function		接口调用结束的回调函数（调用成功、失败都会执行）

示例代码如下。

```
Page({
  onLoad:function(){
    wx.showActionSheet({
      itemList: ['小学', '初中', '高中', '大学'],
      success(res) {
        console.log(res.tapIndex)
      },
      fail(res) {
        console.log(res.errMsg)
      }
    })
  }
})
```

效果如图 6.30 所示。

图 6.30　操作菜单

6.3　定时器 API

微信小程序提供了定时器 API 接口，可以用两种方式设置定时器：setTimeout 在定时到期以后执行注册的回调函数；setInterval 按照指定的周期（以毫秒计）来执行注册的回调函数。这两种方式的区别在于：setTimeout 是达到设定时间后执行一次，如设置 5 分钟后执行定时，那么在 5 分钟后这个定时器就启动执行一次；setInterval 是按照设定的周期来执行，如设置每 5 分钟执行一次，那么每间隔 5 分钟定时器就会执行一次。下面是定时器用到的 API。

慕课视频

定时器 API

（1）number setTimeout（function callback, number delay, any rest）：设定一个定时器，在定时到期以后执行注册的回调函数。

参数说明：function callback 为回调函数；number delay 为延迟的时间，函数的调用会在该延迟之后发生，单位为毫秒；any rest 为附加参数，会作为参数传递给回调函数；返回值为 number 定时器的编号。这个值可以传递给 clearTimeout 来取消该定时。

（2）clearTimeout（number timeoutID）：取消由 setTimeout 设置的定时器。

参数说明：number timeoutID 为要取消的定时器的 ID。

（3）number setInterval（function callback, number delay, any rest）：设定一个定时器，按照

指定的周期（以毫秒计）来执行注册的回调函数。

参数说明：function callback 为回调函数；number delay 为两次执行回调函数之间的时间间隔，单位为毫秒；any rest 为附加参数，会作为参数传递给回调函数；返回值为 number 定时器的编号。这个值可以传递给 clearInterval 来取消该定时。

（4）clearInterval（number intervalID）：取消由 setInterval 设置的定时器。

参数说明：number intervalID 为要取消的定时器的 ID。

示例代码如下。

```
Page({
  onLoad:function(){
    var number1 = setTimeout(this.initTimeout, 1000,  { "name": "小刚" });//启动 setTimeout 定时器
    var number2 = setInterval(this.initInterval, 1000,  { "name": "小刚" });//启动 setInterval 定时器
    //clearTimeout(number1);//取消由 setTimeout 设置的定时器
    //clearInterval(number2);//取消由 setInterval 设置的定时器
  },
  initTimeout:function(res){
    console.log('setTimeout 定时器启动------------------');
    console.log('setTimeout 定时器返回值如下:');
    console.log(res);



  },
  initInterval:function(res){
    console.log('setInterval 定时器启动------------------');
    console.log('setInterval 定时器返回值如下:');
    console.log(res);


  }
})
```

打印日志如图 6.31 所示。

图 6.31　定时器打印日志

6.4　数据缓存 API 的获取

微信小程序提供数据本地缓存功能，如可以将用户信息缓存到本地保存起来，这样就不用每次调用服务器来获取这些信息了。数据缓存 API 就是用来将这些数据保存到本地的，另外还可以获取本地缓存数据、移除缓存数据及清理缓存数据。

数据缓存 API 的获取

6.4.1 将数据缓存到本地

微信小程序为数据缓存到本地提供了两种方式：wx.setStorage（OBJECT）以异步方式将数据存储在本地缓存指定的 key 中；wx.setStorageSync（KEY，DATA）以同步方式将数据存储在本地缓存指定的 key 中。除非用户主动删除或因存储空间原因被系统清理，否则缓存的数据就一直可用。单个 key 允许存储的最大数据长度为 1 MB，所有数据的存储上限为 10 MB。

1. wx.setStorage（OBJECT）

以异步方式将数据存储在本地缓存指定的 key 中，会覆盖掉原来该 key 对应的内容，参数说明如表 6.19 所示。

表 6.19 wx. setStorage 的参数说明

属性	类型	是否必填	说明
key	string	是	本地缓存中指定的 key
data	Object/string	是	需要存储的内容，只支持原生类型、Date 及能够通过 JSON.stringify 序列化的对象
success	Function	否	接口调用成功的回调函数
fail	Function	否	接口调用失败的回调函数
complete	Function	否	接口调用结束的回调函数（调用成功、失败都会执行）

把用户信息缓存到本地的示例代码如下。

```
Page({
  onLoad: function () {
    var user = this.getUserInfo();
    console.log(user);
    wx.setStorage({
      key: 'user',
      data: user,
      success: function (res) {
        console.log(res);
      }
    })
  },
  getUserInfo: function () {
    var user = new Object();
    user.id='10000'
    user.name = 'xiaogang';
    user.userName = '小刚';
    return user;
  }
})
```

在 Storage 里可以查看到缓存的数据，如图 6.32 所示。

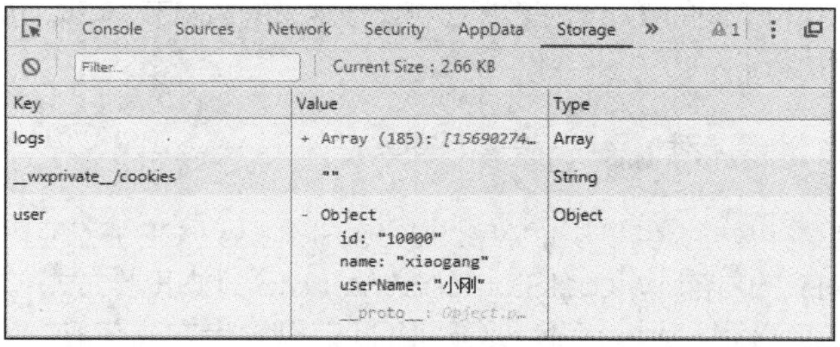

图 6.32 本地缓存数据

2. wx.setStorageSync（KEY，DATA）

以同步方式将数据存储到本地缓存指定的 key 中，会覆盖掉原来该 key 对应的内容，比异步缓存数据更简练一些，参数说明如表 6.20 所示。

表 6.20　wx. setStorageSync 的参数说明

属性	类型	是否必填	说明
key	string	是	本地缓存中指定的 key
data	Object/string	是	需要存储的内容，只支持原生类型、Date 及能够通过 JSON.stringify 序列化的对象

示例代码如下。

```
Page({
  onLoad: function () {
    var userSync = this.getUserInfo();
    console.log(userSync);
    //以同步方式将数据存储到本地
    wx.setStorageSync('userSync', userSync);
  },
  getUserInfo: function () {
    var user = new Object();
    user.id='10000'
    user.name = 'xiaogang';
    user.userName = '小刚'
    return user;
  }
})
```

在 Storage 里可以查看到缓存的数据，如图 6.33 所示。

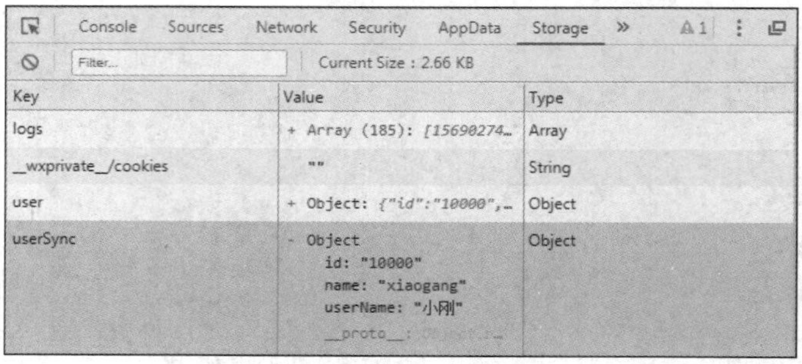

图 6.33　本地缓存数据

将数据缓存到本地，不管是以同步方式还是以异步方式，都是通过 key/value 的形式存储数据的，只不过以同步方式需要等本地缓存成功后，才可以继续执行下面的程序，而以异步方式则不需要等待本地缓存成功就可以继续执行下面的程序。在数据缓存比较耗时的情况下，可以使用异步方式进行缓存，保证程序不用等待继续执行。

6.4.2　获取本地缓存数据

小程序为获取本地缓存数据提供了 4 个 API：wx.getStorage（OBJECT）以异步方式从本地缓存中异步获取指定 key 对应的内容；wx.getStorageSync（KEY）以同步方式从本地缓存中同步获取指定 key 对应的内容；wx.getStorageInfo（OBJECT）以异步方式获取本地所有 key 值集合；wx.getStorageInfoSync 以同步方式获取本地所有 key 值集合。前两个 API 是从指定 key 值里获得缓存数据，而后面两种是获取本地所有 key 值集合。

1. wx.getStorage（OBJECT）

wx.getStorage（OBJECT）使用异步方式从本地缓存中获取指定 key 对应的内容。

参数说明：key 为本地缓存中指定的 key。

在 6.4.1 小节中，我们使用 wx.setStorage（OBJECT），将 user 以异步方式保存到了本地，下面使用 wx.getStorage（OBJECT）来获取本地数据，具体代码如下。

```
Page({
  onLoad:function(){
    //以异步方式获取本地数据
    wx.getStorage({
      key: 'user',
      success: function(res){
        console.log(res);
      }
    })
  }
})
```

获取到的本地数据如图 6.34 所示。

图 6.34　异步获取本地数据

2. wx.getStorageSync（OBJECT）

wx.getStorageSync（OBJECT）是一个同步的接口，用来从本地缓存中同步获取指定 key 对应的内容。

参数说明：key 为本地缓存中指定的 key。

示例代码如下。

```
Page({
  onLoad:function(){
    //以同步方式获取本地数据
    var userSync = wx.getStorageSync('userSync');
    console.log(userSync);
  }
})
```

获取到的本地数据如图 6.35 所示。

图 6.35　同步获取本地数据

3. wx.getStorageInfo（OBJECT）

wx.getStorage 和 wx.getStorageSync 这两个接口都是从本地的指定 key 值来获取数据的，而 wx.getStorageInfo 是以异步方式获取 key 值集合，是获取所有 key 的值，返回值的参数说明如表 6.21 所示。

表 6.21 返回值的参数说明

参数	类型	说明
keys	string Array	当前 storage 中所有的 key
currentSize	number	当前占用的空间大小，单位为千字节
limitSize	number	限制的空间大小，单位为千字节

示例代码如下。

```
Page({
  onLoad:function(){
    wx.getStorageInfo({
      success: function(res){
        console.log(res);
      }
    })
  }
})
```

返回值信息如图 6.36 所示。

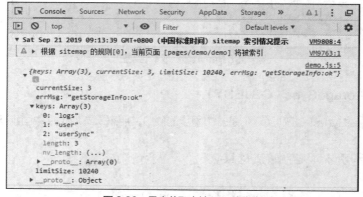

图 6.36 异步获取本地 key 值集合

获取到本地所有的 key 值后，根据该 key 值再调用 wx.getStorage 或 wx.getStorageSync 接口就可以获取到本地数据了。

4. wx.getStorageInfoSync（OBJECT）

wx.getStorageInfoSync 是以同步方法来获取当前 storage 的相关信息，示例代码如下。

```
Page({
  onLoad:function(){
    var storage = wx.getStorageInfoSync();
    console.log(storage);
  }
})
```

返回值信息如图 6.37 所示。

图 6.37 同步获取本地 key 值集合

它和 wx.getStorageInfo 异步获取 storage 返回的数据一样都是所有的 key 值，然后根据 key 值再查找完整的数据。

6.4.3 清理本地缓存数据

wx.removeStorage（OBJECT）、wx.removeStorageSync（KEY）用来从本地缓存中移除指定 key；wx.clearStorage()、wx.clearStorageSync()用来清理本地数据缓存。

1. wx.removeStorage（OBJECT）

wx.removeStorage（OBJECT）用来异步从本地缓存中移除指定的 key。

在未清理前可以看到本地缓存中有 key=user 的数据，如图 6.38 所示。

图 6.38 user 缓存数据

下面从本地缓存中清理 key=user 的数据，具体代码如下。

```
Page({
  onLoad:function(){
    //异步移除 key=user 的数据
    wx.removeStorage({
      key: 'user',
      success: function(res){
        console.log(res);
      },
    })
  }
})
```

清理完后，在本地缓存列表里就找不到 key=user 的缓存数据了，清理成功。

2. wx.removeStorageSync（KEY）

wx.removeStorageSync（OBJECT）用来同步从本地缓存中移除指定的 key，它的效果和 wx.removeStorage 一样。

示例代码如下。

```
Page({
    onLoad:function(){
        //同步移除 key=userSync 的数据
        wx.removeStorageSync('userSync');
    }
})
```

3. wx.clearStorage ()和 wx.clearStorageSync ()

wx. clearStorage ()和 wx. clearStorageSync ()用来清理本地所有缓存数据，前者是以异步方式，后者是以同步方式。

示例代码如下。

```
wx.clearStorage()

try {
    wx.clearStorageSync()
} catch(e) {
}
```

6.4.4 从缓存获取图书列表数据

在莫凡商城的 index.js 文件里，有 getBookList()函数用来获取图书列表数据。下面使用缓存功能将数据缓存起来，并获取缓存的图书列表数据。

改写 index.js 文件中 getBookList()函数的具体代码如下。

```
getBookList: function () {//获取图书列表的方法
    var page = this;
    wx.request({
        url: host + '/api/goods/getHomeGoodsList',
        method: 'GET',
        data: {},
        header: {
            'Content-Type': 'application/json'
        },
        success: function (res) {
            var book = res.data.data;
            //将图书列表数据缓存到本地
            wx.setStorage({
                key: 'book',
                data: book,
            })
            //获取缓存到本地图书列表数据
            book = wx.getStorageSync("book");
            console.log(book);
            var hotList = book.rmjs;//热门书籍列表
            var spikeList = book.mssk;//秒杀时刻书籍列表
            var bestSellerList = book.cxsj;//畅销书籍列表
            page.setData({ hotList: hotList });
            page.setData({ spikeList: spikeList });
            page.setData({ bestSellerList: bestSellerList });
        }
    })
}
```

在 Storage 里可以查看到缓存的数据，如图 6.39 所示。

第 6 章
莫凡商城的注册、登录功能

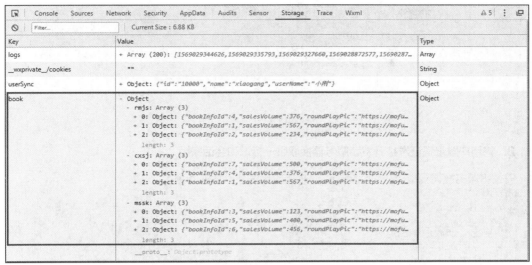

图 6.39 图书列表本地缓存

6.5 登录相关 API

微信小程序登录需要使用登录相关 API、获取账号信息 API、获取用户信息 API、授权 API 及设置 API 接口，我们可以通过这些 API 接口来实现登录、授权、获取用户信息等功能。

6.5.1 登录 API

登录功能是 App 必不可少的功能，微信小程序也提供了登录 API。微信小程序的登录可以简单地理解为以下几个步骤。

（1）在微信小程序里使用 wx.login 方法获取登录凭证 code 值。

（2）将 code 值和 AppId、secret（在公众开发平台中 AppId 的下面）、grant_type 授权类型这 4 个参数发送到自己后台开发的服务器上，在后台服务器上去请求路径 https://api.weixin.qq.com/sns/jscode2session（该链接是微信请求的接口地址，直接打开不能使用，需要传递以上参数才能使用）。同时传递这 4 个参数，就能获取唯一标识（openid）和会话密钥（session_key）。

（3）拿到唯一标识和会话密钥后，在自己后台开发的服务器上生成自己的 sessionId。

（4）微信小程序可以将服务器生成的 sessionId 信息保存到本地缓存（Storage）。

（5）后续用户进入微信小程序，先从本地缓存获得 sessionId，将这个 sessionId 传输到服务器上进行查询来维护登录状态。

下面一起来看一下这些步骤的实现。

1．用 wx.login（OBJECT）获取登录凭证 code

微信小程序使用 wx.login 接口来获取登录凭证（code），用户允许登录后，回调内容会带上 code（有效时间为 5 分钟），开发者需要将 code 发送到开发者服务器后台，来获取 openid 和 session_key。

示例代码如下。

```
App({
  onLaunch: function() {
    wx.login({
      success:function(res){
```

```
            var code = res.code; //用户登录凭证
            if(code){
                console.log('获取用户登录凭证:'+code);
            }else{
                console.log('获取用户登录凭证失败');
            }
        }
    })
    }
})
```

2. 将用户登录凭证发往开发者服务器换取唯一标识和会话密钥

开发者服务器需要提供一个后台接口，来接收用户登录凭证 code。

```
App({
    onLaunch: function() {
        wx.login({
            success:function(res){
                var code = res.code; //用户登录凭证
                if(code){
                    console.log('获取用户登录凭证:'+code);
                    wx.request({ //请求自己的后台服务器，传输用户登录凭证 code
                        url: 'https://www.my-domain.com/wx/onlogin',
                        data: { code: code }
                    })
                }else{
                    console.log('获取用户登录凭证失败');
                }
            }
        })
    }
})
```

3. 开发者服务器使用 code 值获取唯一标识和会话密钥

开发者服务器接收到用户登录凭证（code）值后，用其与小程序唯一标识 appid、小程序 secret、小程序的授权类型 grant_type 这 4 个参数去请求微信服务器接口 https://api.weixin.qq.com/sns/jscode2session，来获取会话密钥 session_key 和唯一标识 openid。其中，session_key 是对用户数据进行加密签名的密钥。为了自身的应用安全，session_key 不应该在网络上传输。

接口地址为：https://api.weixin.qq.com/sns/jscode2session?appid=APPID&secret=SECRET&js_code=JSCODE&grant_type=authorization_code。

参数说明：appid 为小程序唯一标识；secret 为小程序的 App secret；js_code 为登录时获取的 code；grant_type 填写为"authorization_code"固定值。

返回值说明：openid 为用户在微信里的唯一标识；session_key 为会话密钥。

后台服务器请求微信服务器代码的实现可以使用 Java、PHP、C++、Node.js 等多种语言。

4. 开发者服务器生成自己的 sessionId

开发者服务器获取到唯一标识（openid）和会话密钥（session_key）后，需要生成自己的 sessionId，规则可以由自己制定，可以拼接成字符串，也可以拼接成字符串后再用 MD5 加密等。生成的 sessionId 需要在开发者服务器中保存起来，小程序在校验登录或者进行需要登录后才能做的操作时，都需要到开发者服务器中来验证。sessionId 可以保存到缓存 Memcached、Redis 或内存中。

5. 小程序客户端保存 sessionId

小程序客户端是没有类似于浏览器客户端的 cookie 或者 session 机制的，但是可以利用小程序的 storage 缓存机制来保存 sessionId，在需要登录状态才能发起请求的时候传递这个参数，从而不用每次重新登录。在之后调用那些需要登录后才有权限访问的后台服务时，可以将保存在 storage 中的 sessionId 取出并携带在请求中，传递到后台服务，后台代码获取到该 sessionId 后，从缓存 Redis 或者内存中查找是否有该 sessionId 存在，存在即确认该 session 是有效的，继续执行后续的代码，否则进行错误处理。

6. wx.checkSession（OBJECT）检查登录态是否过期

微信小程序可以使用 wx.checkSession（OBJECT）来检查登录态是否过期，如果过期就重新登录。

通过 wx.login 接口获得的用户登录态拥有一定的时效性。用户越久未使用小程序，其登录态就越有可能失效；反之，如果用户一直在使用小程序，则其登录态将一直保持有效。具体时效逻辑由微信维护，对开发者透明。开发者只需要调用 wx.checkSession 接口检测当前用户登录态是否有效即可。

登录态过期后，开发者可以再调用 wx.login 获取新的用户登录态。调用成功说明当前 session_key 未过期；调用失败则说明 session_key 已过期。示例代码如下。

```
wx.checkSession({
    success: function(){
        //登录态未过期
    },
    fail: function(){
        //登录态过期
        wx.login()
    }
})
```

6.5.2 获取账号信息 API

微信小程序提供 Object wx.getAccountInfoSync() API 来获取账号信息。账号信息包括两方面内容：miniProgram 小程序账号信息和 plugin 插件账号信息（仅在插件中调用时包含这一项）。miniProgram 小程序账号信息里包括小程序 appId；plugin 插件账号信息里包括插件 appId 和 version 插件版本号。

示例代码如下。

```
const accountInfo = wx.getAccountInfoSync();
console.log(accountInfo.miniProgram.appId) // 小程序 appId
console.log(accountInfo.plugin.appId) // 插件 appId
console.log(accountInfo.plugin.version) // 插件版本号, 'a.b.c' 这样的形式
```

6.5.3 获取用户信息 API

微信小程序使用 wx.getUserInfo（OBJECT）来获取用户信息，在获取用户信息之前，需要调用 wx.login 接口。只有用户在登录状态，才能获取到用户的相关信息；在用户未授权过的情况下调用此接口，将不再出现授权弹窗。具体参数如表 6.22 所示。

表 6.22 wx. getUserInfo 的参数说明

属性	类型	是否必填/默认值	说明
withCredentials	boolean	否	是否带上登录态信息。当 withCredentials 为 true 时，要求此前调用过 wx.login 且登录态尚未过期，此时返回的数据会包含 encryptedData、iv 等敏感信息，如表 6.23 所示；当 withCredentials 为 false 时，不要求有登录态，返回的数据不包含 encryptedData、iv 等敏感信息

续表

属性	类型	是否必填/默认值	说明
lang	string	en	显示用户信息的语言，en 为英文、zh_CN 为简体中文、zh_TW 为繁体中文
success	Function	否	接口调用成功的回调函数，参数说明如表 6.23 所示
fail	Function	否	接口调用失败的回调函数
complete	Function	否	接口调用结束的回调函数（调用成功、失败都会执行）

表 6.23　success 返回参数说明

属性	类型	说明
userInfo	UserInfo	用户信息对象，包括 nickName 用户昵称、avatarUrl 用户头像图片的 URL、gender 用户性别（0 为未知、1 为男性、2 为女性）、country 用户所在国家、province 用户所在省份、city 用户所在城市、language 显示 country、province、city 所用的语言（en 为英文、zh_CN 为简体中文、zh_TW 为繁体中文）
rawData	string	不包括敏感信息的原始数据字符串，用于计算签名
signature	string	使用 sha1（rawData+sessionkey）得到字符串，用于校验用户信息
encryptedData	string	包括敏感数据在内的完整用户信息的加密数据
iv	string	加密算法的初始向量
cloudID	string	敏感数据对应的云 ID，开通云开发的小程序才会返回，可通过云调用直接获取开放数据

示例代码如下。

```
Page({
  onLoad: function () {
    // 必须是在用户已经授权的情况下调用
    wx.getUserInfo({
      success: function (res) {
        var userInfo = res.userInfo
        var nickName = userInfo.nickName
        var avatarUrl = userInfo.avatarUrl
        var gender = userInfo.gender //性别 0:未知、1:男、2:女
        var province = userInfo.province
        var city = userInfo.city
        var country = userInfo.country
      }
    })
  }
})
```

6.5.4　授权 API

微信小程序部分 API 接口需要通过 wx.authorize（Object object）来向用户发起授权请求，调用后会立刻弹窗询问用户是否同意授权小程序使用某项功能或获取用户的某些数据，但不会实际调用对应接口。如果用户之前已经同意授权，则不会出现弹窗，直接返回成功。开发者可以使用 wx.getSetting 获取用户当前的授权状态。

打开设置界面，用户可以在小程序设置界面中控制对该小程序的授权状态，如图 6.40 所示。开发者可以调用 wx.openSetting 打开设置界面，引导用户开启授权。

打开小程序设置页将支持以下两种实现方式。

（1）使用 button 组件来使用此功能，示例代码如下。

```
<button open-type="openSetting" bindopensetting="callback">打开设置页</button>
```

（2）由单击行为触发 wx.openSetting 接口的调用，示例代码如下。

```
<button bindtap="openSetting">打开设置页</button>
Page({
```

```
openSetting:function(){
  wx.openSetting()
}
})
```

图 6.40　打开设置界面

wx.authorize（Object object）授权接口通过属性 scope 来设置授权，scope 授权如表 6.24 所示。

表 6.24　scope 授权列表

scope	接口	说明
scope.userInfo	wx.getUserInfo	用户信息
scope.userLocation	wx.getLocation, wx.chooseLocation	地理位置
scope.userLocationBackground	wx.startLocationUpdateBackground	后台定位
scope.address	wx.chooseAddress	通信地址
scope.invoiceTitle	wx.chooseInvoiceTitle	发票抬头
scope.invoice	wx.chooseInvoice	获取发票
scope.werun	wx.getWeRunData	微信运动步数
scope.record	wx.startRecord	录音功能
scope.writePhotosAlbum	wx.saveImageToPhotosAlbum, wx.saveVideoToPhotosAlbum	保存到相册
scope.camera	camera 组件	摄像头

示例代码如下。

```
Page({
  onLoad: function () {
    // 可以通过 wx.getSetting 先查询一下用户是否授权了 "scope.record" 这个 scope
    wx.getSetting({
      success(res) {
        if (!res.authSetting['scope.record']) {
          wx.authorize({
            scope: 'scope.record',
            success() {
              // 用户已经同意小程序使用录音功能，后续调用 wx.startRecord 接口不会弹窗询问
              wx.startRecord()
            }
          })
        }
```

                }
            })
        }
    })
效果如图 6.41 所示。

图 6.41 授权弹窗

6.5.5 设置 API

微信小程序提供的设置相关 API 接口如下。

（1）wx.openSetting（Object object）：打开客户端小程序设置界面，返回用户设置的操作结果，设置界面中只会出现小程序已经向用户请求过的权限。

（2）wx.getSetting（Object object）：获取用户的当前设置，返回值中只会出现小程序已经向用户请求过的权限。

返回授权结果 AuthSetting 对象的属性如下。

（1）boolean scope.userInfo：是否授权用户信息，对应接口 wx.getUserInfo。

（2）boolean scope.userLocation：是否授权地理位置，对应接口 wx.getLocation, wx.chooseLocation。

（3）boolean scope.address：是否授权通信地址，对应接口 wx.chooseAddress。

（4）boolean scope.invoiceTitle：是否授权发票抬头，对应接口 wx.chooseInvoiceTitle。

（5）boolean scope.invoice：是否授权获取发票，对应接口 wx.chooseInvoice。

（6）boolean scope.werun：是否授权微信运动步数，对应接口 wx.getWeRunData。

（7）boolean scope.record：是否授权录音功能，对应接口 wx.startRecord。

（8）boolean scope.writePhotosAlbum：是否授权保存到相册对应接口 wx.saveImageToPhotosAlbum、wx.saveVideoToPhotosAlbum。

（9）boolean scope.camera：是否授权摄像头，对应[camera]（(camera)）组件。

wx.openSetting 从 2.3.0 版本开始，用户发生单击行为后，才可以跳转打开设置页，管理授权信息，示例代码如下。

```
<!--wxml 页面-->
<!--方法 1:使用 button 组件来使用此功能-->
<button open-type="openSetting" bindopensetting='handler'>单击授权并获取位置信息</button>

<!--方法 2:由单击行为触发 wx.openSetting 接口的调用-->
<button bindtap="settingBtn">打开设置页</button>
```

```
Page({
    handler: function (e) {
        var that = this;
        if (!e.detail.authSetting['scope.userLocation']) {
```

```
        //打开设置界面
    }
  },
  settingBtn:function(){
    wx.openSetting();
  }
})
```

wx.getSetting 示例代码如下。

```
Page({
  onLoad: function () {
    //可以通过 wx.getSetting 先查询一下用户是否授权了 "scope.record" 这个 scope
    wx.getSetting({
      success(res) {
        if (!res.authSetting['scope.record']) {
          wx.authorize({
            scope: 'scope.record',
            success() {
              // 用户已经同意小程序使用录音功能，后续调用 wx.startRecord 接口不会弹窗询问
              wx.startRecord()
            }
          })
        }
      }
    });
  }
})
```

6.6 项目实战：任务 3——实现登录功能

慕课视频

项目实战：实现登录功能

1. 任务目标

通过实现莫凡商城的登录功能，学会登录功能要应用到的组件和 API 接口的使用方法，并能举一反三，实现其他类似的登录功能。

莫凡商城登录功能提供两种登录方式：账号密码登录和手机快捷登录。通过页签的切换，可以选择使用哪种方式进行登录，如图 6.42 和图 6.43 所示。

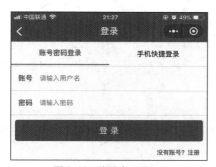

图 6.42　账号密码登录

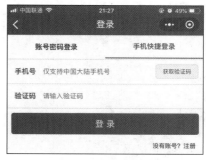

图 6.43　手机快捷登录

2. 任务实施

下面我们一起来实现莫凡商城登录功能。

（1）在 app.json 文件里添加注册页面路径"pages/login/login"。

（2）在 login.json 文件里配置导航标题，示例代码如下。

```
{
    "navigationBarTitleText": "登录"
}
```

（3）在login.wxml页面文件里进行登录表单布局，需要使用view容器组件、form表单组件、swiper滑块视图容器组件和button按钮组件，示例代码如下。

```
<form bindsubmit="formSubmit" bindreset="formReset">
  <view class="content">
    <view class="loginTitle">
      <view class="{{currentTab==0?'select':'default'}}" data-current="0" bindtap="switchNav">账号密码登录</view>
      <view class="{{currentTab==1?'select':'default'}}" data-current="1" bindtap="switchNav">手机快捷登录</view>
    </view>
    <view class="hr"></view>
    <swiper current="{{currentTab}}" style="height:{{winHeight}}px">
      <swiper-item>
        <view class="accountType">
          <view class="account">
            <view class="ac">账号</view>
            <view class="ipt"><input name="loginName" focus="false" placeholder="请输入用户名" class="placeholder-style" value='{{form_info}}'/></view>
          </view>
          <view class="hr"></view>
          <view class="account">
            <view class="ac">密码</view>
            <view class="ipt"><input name="loginPassword" type="text" placeholder="请输入密码" class="placeholder-style" value='{{form_info}}'/></view>
          </view>
          <view class="hr"></view>
          <view class="login">
            <button form-type="submit">登录</button>
            <view class="fp" bindtap='toRegister'>没有账号？注册</view>
            <view class="tip">{{tip}}</view>
          </view>
        </view>
      </swiper-item>
      <swiper-item>
        <view class="mobileType">
          <view class="account">
            <view class="ac">手机号</view>
            <view class="ipt"><input name="mobile" type="text" placeholder="仅支持中国大陆手机号" class="placeholder-style" value='{{form_info}}' bindinput='getMobile'/>
              <button class="btn" bindtap="getcode" wx:if="{{flag==true}}">{{yzmvalue}}</button>
              <button class="btn" wx:else>{{timevalue}}s</button>
            </view>
          </view>
          <view class="hr"></view>
          <view class="account">
            <view class="ac">验证码</view>
            <view class="ipt"><input name="verifyCode" type="text" placeholder="请输入验证码" class="placeholder-style" value='{{verifyCode}}'/></view>
          </view>
          <view class="hr"></view>
          <view class="login">
            <button form-type="submit">登录</button>
            <view class="fp" bindtap='toRegister'>没有账号？注册</view>
            <view class="tip">{{tip}}</view>
          </view>
        </view>
      </swiper-item>
    </swiper>
  </view>
</form>
```

（4）在 login.wxss 样式文件里对登录表单布局进行样式渲染，示例代码如下。

```css
.loginTitle{
    display: flex;
    flex-direction: row;
    width: 100%;
    font-size: 15px;
}
.select{
    color: #009966;
    width: 50%;
    text-align: center;
    height: 48px;
    line-height: 48px;
    border-bottom:5rpx solid #009966;
    font-weight: bold;
}
.default{
    margin: 0 auto;
    padding: 15px;
}
.hr{
    height: 1px;
    width: 100%;
    background-color: #666666;
    opacity: 0.2;
}
.account{
    display: flex;
    flex-direction: row;
}
.ac{
    padding:15px;
    font-size:15px;
    font-weight: bold;
    color: #666666;
}
.ipt{
    padding-top:10px;
}
.ipt input{
    text-align: left;
    width: 200px;
    color: #000000;
}
.placeholder-style{
    font-size: 14px;
    color: #cccccc;
}
.login{
    margin: 0 auto;
    text-align: center;
    padding-top:10px;
}
.login button{
   width: 96%;
   color: #ffffff;
   background: #009966;
}
.fp{
    font-size: 13px;
```

```
      color: #3e13da;
      padding:5px;
      text-align: right;
      margin-right:10px;
      margin-top:10px;
}
.btn{
      position: absolute;
      right: 10px;
      top: 10px;
      width: 90px;
      font-size: 12px;
      color: #666666;
      background-color: #f2f2f2;
}
.tip{
      margin-top:10px;
      font-size: 12px;
      color: #D53E37;
}
```

（5）在 login.js 业务逻辑处理文件中，进行登录表单切换、登录表单验证，然后将登录表单提交到后台服务器中进行登录，示例代码如下。

```
var app = getApp();
var host = app.globalData.host;
var timer;
var timeSecond = false, sendBolen = false;
Page({
  data: {
    currentTab: 0,
    winWidth: 0,
    winHeight: 0,
    tip: '',
    form_info: '',
    yzmvalue: '获取验证码',
    mobile: '',
    timevalue: 60,
    flag: true,
    verifyCode: ''
  },
  onLoad: function (options) {
    var userId = wx.getStorageSync("userId");
    if (userId == "") {
      this.checklogin();
    } else {
      wx.reLaunch({
        url: '../index/index',
      })
    }

    var page = this;
    wx.getSystemInfo({
      success: function (res) {
        console.log(res);
        page.setData({ winWidth: res.windowWidth });
        page.setData({ winHeight: res.windowHeight });
      }
    })
  },
  switchNav: function (e) {
    var that = this;
```

```js
      if (this.data.currentTab == e.target.dataset.current) {
        return false;
      } else {
        that.setData({ currentTab: e.target.dataset.current });
        that.setData({ tip: '' });
        that.setData({ form_info: '' });
      }
    },
    toRegister: function (e) {
      wx.navigateTo({
        url: '../register/register'
      })
    },
    formSubmit: function (e) {
      var that = this;
      var loginName = e.detail.value.loginName;
      var mobile = e.detail.value.mobile;
      var loginPassword = e.detail.value.loginPassword;
      var verifyCode = e.detail.value.verifyCode;
      var loginType = that.data.currentTab;
      var code = app.globalData.code;
      //验证表单输入
      var ret = that.checkLogin(loginName, mobile, loginPassword, verifyCode, loginType);
      if (ret) {
        wx.request({
          url: host + '/api/user/swxLogin',
          method: 'GET',
          data: { 'loginName': loginName, 'mobile': mobile, 'loginPassword': loginPassword, 'verifyCode': verifyCode, 'loginType': loginType, 'code': code },
          header: {
            'Content-Type': 'application/json'
          },
          success: function (res) {
            console.log(res);
            var code = res.data.code;
            var msg = res.data.data;
            if (code == '0000') {
              app.globalData.userId = res.data.data.user.userId;
              wx.setStorageSync('userId', res.data.data.user.id);
              wx.setStorageSync('nickName', res.data.data.user.nickName)
              wx.setStorageSync('swx_session', res.data.data.swx_session);
              wx.setStorageSync('userMobile', res.data.data.user.mobile);
              wx.setStorageSync('openId', res.data.data.openId);
              wx.setStorageSync('token', res.data.data.token);
              wx.reLaunch({
                url: '../index/index'
              })
              console.log("2")
            } else {
              that.setData({ tip: msg });
              return false
            }
          }
        })
      }
    },
    checkLogin: function (loginName, mobile, loginPassword, verifyCode, loginType) {
      var that = this;
      if (loginType == 0) {
        if (loginName == "") {
```

```
          that.setData({ tip: '用户名不能为空！' });
          return false
        }
        if (loginPassword == '') {
          that.setData({ tip: '密码不能为空！' });
          return false
        }
      } else {
        if (mobile == '') {
          that.setData({ tip: '手机号不能为空！' });
          return false
        }

        var myreg = /^[1][3, 4, 5, 7, 8][0-9]{9}$/;
        if (!myreg.test(mobile)) {
          that.setData({ tip: '手机号不合法！' });
          return false;
        }

        if (verifyCode == '') {
          that.setData({ tip: '验证码不能为空！' });
          return false
        }
      }

      that.setData({ tip: '' });
      return true
    },
    checklogin: function () {//登录页面进行一次获取新的 code
      wx.login({
        success: function (data) {
          console.log(data)
          app.globalData.code = data.code;
        }
      })
    },
    getcode: function (e) {
      var that = this;
      if (that.data.mobile == "") {
        that.setData({ tip: '请输入手机号' });
        return false;
      }
      var myreg = /^[1][3, 4, 5, 7, 8][0-9]{9}$/;
      if (!myreg.test(that.data.mobile)) {
        that.setData({ tip: '手机号不合法！' });
        return false;
      }
      that.setData({ tip: '' });//去除提示
      that.setData({ flag: false });//显示时间
      timer = setInterval(that.settime, 1000);//验证码倒计时
      wx.request({
        url: host + '/api/user/getVerifyCode',
        method: 'GET',
        data: {
          'mobile': this.data.mobile
        },
        header: {
          'Content-Type': 'application/json'
        },
        success: function (res) {
```

```
          console.log(res);
          var code = res.data.code;
          var msg = res.data.data;
          if (code == '0000') {
            clearInterval(timer);
            that.setData({ verifyCode: msg });
          } else {
            clearInterval(timer);
            that.setData({
              yzmvalue: '获取验证码',
              timevalue: 60,
              flag: true
            });
            that.setData({ tip: msg });
            return false
          }
        }
      })
    },
    getMobile: function (e) {
      this.setData({
        mobile: e.detail.value
      })
    },
    settime: function () {
      var timevalue = this.data.timevalue;

      if (timevalue == 0) {
        clearInterval(timer)
        this.setData({
          yzmvalue: '重新获取',
          timevalue: 60,
          flag: true
        })
        timeSecond = false;
        sendBolen = false;
        return;
      }
      timevalue--;
      timeSecond = true;
      sendBolen = true;
      this.setData({
        timevalue: timevalue,
        flag: false
      })
    },
  })
```

登录功能是很常用的功能，我们可以通过莫凡商城项目学会登录功能的设计及实现方法，通过综合应用 view 容器组件、form 表单组件、swiper 滑块视图容器组件、button 按钮组件进一步理解组件的使用；同时，我们还应用了 wx.request 网络请求 API、setInterval 定时器 API 及缓存相关 API 接口，这些都是常用的功能。

6.7 项目实战：任务 4——实现"我的"界面列表式导航功能（2）

1. 任务目标

通过实现莫凡商城"我的"界面列表式导航功能，学会列表式导航设计的方式。很多 App 都会采用列表式导航设计，本任务的实现对其他类似项目的设计都有借鉴作用。

莫凡商城"我的"界面包括账号登录区域、我的订单区域、列表式导航区域，如图6.44所示。

图6.44 "我的"界面

慕课视频

项目实战：实现"我的"界面列表式导航功能（2）

2. 任务实施

下面我们一起来实现莫凡商城"我的"界面列表式导航功能。

（1）在app.json文件里添加我的页面路径"pages/me/me"。

（2）在me.wxml页面文件里进行账号区域布局设计、我的订单区域布局设计、列表式导航布局设计，示例代码如下。

```
<view class="content">
  <view class="head">
    <view class="headIcon"><image src="/pages/images/icon/head.jpg" style="width:70px;height:70px;"></image></view>
    <view class="login"><navigator url="../login/login" hover-class="navigator-hover">{{nickName}}</navigator></view>
    <view class="detail"><text></text></view>
  </view>
  <view class="hr"></view>
  <view style="display:flex;flex-direction:row;">
    <view class="order">我的订单</view>
    <view class="detail2"><text></text></view>
  </view>
  <view class="line"></view>
  <view class="nav">
    <view class="nav-item" bindtap="nav" id="0" data-status='1'>
      <view><image src="/pages/images/icon/dfk.png" style="width:28px;height:25px;"></image></view>
      <view>待付款</view>
    </view>
    <view class="nav-item" bindtap="nav" id="1" data-status='3'>
      <view><image src="/pages/images/icon/dsh.png" style="width:36px;height:27px;"></image></view>
      <view>待收货</view>
    </view>
    <view class="nav-item" bindtap="nav" id="2" data-status='4'>
      <view><image src="/pages/images/icon/dpj.png" style="width:31px;height:28px;"></image></view>
      <view>已完成</view>
    </view>
```

```
      </view>
      <view class="hr"></view>

      <view class="item">
        <view class="order">我的消息</view>
        <view class="detail2"><text></text></view>
      </view>
      <view class="line"></view>
      <view class="item">
        <view class="order">我的收藏</view>
        <view class="detail2"><text></text></view>
      </view>
       <view class="line"></view>
      <view class="item">
        <view class="order">账户余额</view>
        <view class="detail2"><text>0.00 元 </text></view>
      </view>
      <view class="line"></view>
     <view class="hr"></view>
      <view class="item" bindtap="updatePwd">
          <view class="order">修改密码</view>
          <view class="detail2"><text></text></view>
      </view>
      <view class="line"></view>
      <view class="item" bindtap="opinion">
          <view class="order">意见反馈</view>
          <view class="detail2"><text></text></view>
      </view>
       <view class="line"></view>
      <view class="item" bindtap='clearStore'>
          <view class="order" >清除缓存</view>
          <view class="detail2"><text ></text></view>
      </view>
      <view class="line"></view>
     <view class="hr"></view>
       <view class="line"></view>
      <view class="item">
          <view class="order">知识扩展</view>
          <view class="detail2"><text ></text></view>
      </view>
      <view class="hr"></view>
</view>
```

（3）在 me.wxss 样式文件里进行账号区域样式渲染、我的订单区域样式渲染、列表式导航样式渲染，示例代码如下。

```
.head{
    width:100%;
    height: 90px;
    background-color: #009966;
    display: flex;
    flex-direction: row;
}
.headIcon{
    margin: 10px;
}
.headIcon image{
    border-radius:50%;
}
.login{
    color: #ffffff;
    font-size: 15px;
```

```css
        font-weight: bold;
        position: absolute;
        left:100px;
        margin-top:30px;
    }
    .detail{
        color: #ffffff;
        font-size: 15px;
        position: absolute;
        right: 10px;
        margin-top: 30px;
    }
    .nav{
        display: flex;
        flex-direction: row;
        padding-top:10px;
        padding-bottom: 10px;
    }
    .nav-item{
        width: 25%;
        font-size: 13px;
        text-align: center;
        margin:0 auto;
    }
    .hr{
        width: 100%;
        height: 15px;
        background-color: #f5f5f5;
    }
    .order{
        padding-top:15px;
        padding-left: 15px;
        padding-bottom:15px;
        font-size:15px;
    }
    .detail2{
        font-size: 15px;
        position: absolute;
        right: 10px;
        margin-top:15px;
        color: #888888;
    }
    .line{
         height: 1px;
        width: 100%;
        background-color: #666666;
        opacity: 0.2;
    }
    .item{
      display:flex;
      flex-direction:row;
    }
```

（4）在 me.js 业务逻辑处理文件里，添加昵称"nickName"，默认值为"立即登录"，设置"我的"界面标题，校验是否登录，如果没有登录，则跳转到登录界面，示例代码如下。

```javascript
var app = getApp();
var host = app.globalData.host;
Page({
  data: {
    nickName: '立即登录'
  },
```

```
onLoad: function(options) {
  this.checkLogin();//校验是否登录
  wx.setNavigationBarTitle({//设置标题
    title: '我的'
  })
},
checkLogin: function() {//校验是否登录
  var userId = wx.getStorageSync("userId");
  if (userId == null || userId == "") {
    wx.navigateTo({
      url: '../login/login',
    })
  } else {
    this.setData({
      nickName: wx.getStorageSync("nickName")
    });
  }
}
})
```

通过实现"我的"界面列表式导航功能，我们要学会处理用户登录状态，没有登录的时候，显示立即登录，登录后显示昵称，这是一种常规的处理方式；同时，要学会对列表式导航进行布局，这也是"我的"界面中常用的一种设计方式。

6.8 项目实战：任务 5——实现修改密码功能

1. 任务目标

通过实现莫凡商城的修改密码功能，我们可以学会通过表单组件来完成修改密码界面的设计，以及进行表单校验，实现密码修改功能。

莫凡商城修改密码界面包括原密码、新密码、确认密码 3 个模块，如图 6.45 所示。

图 6.45　修改密码

2. 任务实施

下面我们一起来实现莫凡商城修改密码的功能。

（1）在 app.json 文件里添加修改密码页面路径"pages/updatePwd/updatePwd"。

（2）在"我的"界面列表式导航菜单修改密码处添加绑定事件，这样单击修改密码菜单时，就会跳转到修改密码页面，示例代码如下。

```
var app = getApp();
var host = app.globalData.host;
Page({
  data: {
```

```
      nickName: '立即登录'
    },
    onLoad: function(options) {
      this.checkLogin();//校验是否登录
      wx.setNavigationBarTitle({//设置标题
        title: '我的'
      })
    },
    checkLogin: function() {//校验是否登录
      var userId = wx.getStorageSync("userId");
      if (userId == null || userId == "") {
        wx.navigateTo({
          url: '../login/login',
        })
      } else {
        this.setData({
          nickName: wx.getStorageSync("nickName")
        });
      }
    },
    nav: function(e) {//我的订单跳转
      var id = e.currentTarget.id;
      var status = e.currentTarget.dataset.status;
      wx.navigateTo({
        url: '../myOrder/myOrder?id=' + id + '&status=' + status
      })
    },
    updatePwd: function(e) {//修改密码
      wx.navigateTo({
        url: '../updatePwd/updatePwd'
      })
    }
})
```

（3）在updatePwd.wxml页面文件里进行修改密码界面布局，示例代码如下。

```
<view class="content">
  <form bindsubmit="formSubmit" bindreset="formReset">
    <view class="items">
      <view class="item">
        <view class="title">原密码</view>
        <view class="input">
          <input name="oldPwd"    type="text" password placeholder="请输入原密码" placeholder-class="holder" />
        </view>
      </view>
      <view class="line"></view>
      <view class="item">
        <view class="title">新密码</view>
        <view class="input">
          <input name="newPwd" type="text" password placeholder="请输入新密码" placeholder-class="holder" />
        </view>
      </view>
      <view class="line"></view>
      <view class="item">
        <view class="title">确认密码</view>
        <view class="input">
          <input name="confirmPwd" type="text" password placeholder="请输入确认密码" placeholder-class="holder" />
        </view>
      </view>
      <view class="line"></view>
    </view>
```

```
        <button class="btn" form-type="submit">提交</button>
    </form>
</view>
```

（4）在 updatePwd.wxss 样式文件里对修改密码界面进行样式渲染，示例代码如下。

```css
.content{
    background-color: #F9F9F8;
    height: 600px;
    font-family: "Microsoft YaHei";
}
.item{
    display: flex;
    flex-direction: row;
    background-color: #ffffff;
    padding: 10px;
    height: 25px;
    line-height: 25px;
}
.title{
    font-size:16px;
    font-weight: bold;
    color: #666666;
    width: 70px;
}
.input{
    margin-left: 10px;
}
.holder{
    font-size: 14px;
    color: #cccccc;
}
.line{
    height: 1px;
    width: 100%;
    background-color: #666666;
    opacity: 0.2;
}
.btn{
    background-color: #009966;
    margin: 10px;
    color:#ffffff;
}
```

（5）在 updatePwd.js 业务逻辑处理文件中进行表单校验和修改密码操作的实现，示例代码如下。

```js
var app = getApp();
var host = app.globalData.host;
Page({
    data: {

    },
    onLoad: function (options) {

    },
    formSubmit: function (e) {//校验表单和提交表单
        var page = this;
        var oldPwd = e.detail.value.oldPwd;
        var newPwd = e.detail.value.newPwd;
        var confirmPwd = e.detail.value.confirmPwd;
        if (oldPwd == null || oldPwd == '') {
            page.showTip("原密码不能为空");
            return;
        }
```

```
            if (newPwd == null || newPwd == '') {
              page.showTip("新密码不能为空");
              return;
            }
            if (confirmPwd == null || confirmPwd == '') {
              page.showTip("确认密码不能为空");
              return;
            }
            if (confirmPwd != newPwd) {
              page.showTip("新密码与确认密码不一致");
              return;
            }

            var userId = wx.getStorageSync("userId");
            if (userId == null || userId == "") {
              wx.navigateTo({
                url: '../login/login',
              })
            } else {
              wx.request({
                url: host + '/api/user/updatePwd',
                method: 'GET',
                data: {
                  userId: userId,
                  oldPwd: oldPwd,
                  newPwd: newPwd
                },
                header: {
                  'Content-Type': 'application/json'
                },
                success: function (res) {
                  var code = res.data.code;
                  if (code == '0000') {
                    wx.showToast({
                      title: '修改成功',
                      icon: 'success',
                      duration: 1000,
                      success: function (res) {
                        wx.reLaunch({
                          url: '../me/me'
                        })
                      }
                    })
                  }
                }
              })
            }
        },
        showTip:function(content){//表单校验弹窗提示
          wx.showModal({
            title: '提示',
            content: content,
            showCancel: false
          });
        }
    })
```

修改密码功能的实现主要在于表单布局设计和表单的校验，校验通过后调用后台服务器接口来修改密码，修改成功后弹窗提示。

6.9 项目实战：任务 6——实现意见反馈功能

1. 任务目标

通过实现莫凡商城的意见反馈功能，我们可以学会设计意见反馈功能，通过用户提交的意见反馈来改善 App 或者小程序。

莫凡商城意见反馈界面包括意见反馈输入框和"提交"按钮，如图 6.46 所示。

图 6.46　意见反馈

2. 任务实施

下面我们一起来实现莫凡商城意见反馈功能。

（1）在 app.json 文件里添加意见反馈界面路径 "pages/opinion/opinion"。

（2）在"我的"界面列表式导航菜单意见反馈处添加绑定事件，这样单击意见反馈菜单时，就会跳转到意见反馈页面，示例代码如下。

```
var app = getApp();
var host = app.globalData.host;
Page({
  data: {
    nickName: '立即登录'
  },
  onLoad: function(options) {
    this.checkLogin();//校验是否登录
    wx.setNavigationBarTitle({//设置标题
      title: '我的'
    })
  },
  checkLogin: function() {//校验是否登录
    var userId = wx.getStorageSync("userId");
    if (userId == null || userId == "") {
      wx.navigateTo({
        url: '../login/login',
      })
    } else {
      this.setData({
        nickName: wx.getStorageSync("nickName")
      });
    }
  },
  nav: function(e) {//我的订单跳转
    var id = e.currentTarget.id;
    var status = e.currentTarget.dataset.status;
    wx.navigateTo({
      url: '../myOrder/myOrder?id=' + id + '&status=' + status
    })
  },
  updatePwd: function(e) {//修改密码
    wx.navigateTo({
      url: '../updatePwd/updatePwd'
    })
  },
  opinion: function (e) {//意见反馈
    wx.navigateTo({
      url: '../opinion/opinion'
    })
```

（3）在 opinion.json 配置文件里配置导航标题为意见反馈，示例代码如下。

```
{
  "navigationBarTitleText": "意见反馈"
}
```

（4）在 opinion.wxml 页面文件里进行意见反馈布局设计，示例代码如下。

```
<view class="content">
  <form bindsubmit="formSubmit" bindreset="formReset">
    <view class="opinion">
      <textarea placeholder="请填写您的意见或建议" placeholder-class='holder' name="content">
      </textarea>
    </view>
    <button   class="btn" form-type="submit">提交</button>
  </form>
</view>
```

（5）在 opinion.wxss 样式文件里对意见反馈样式进行渲染，示例代码如下。

```
.content{
  background-color: #F9F9F8;
  height: 600px;
  font-family: "Microsoft YaHei";
}
.opinion{
  margin:10px;
  background-color:#ffffff;
  border-radius: 5px;
}
.holder{
  font-size: 13px;
  color: #999999;
}
.btn{
  background-color: #009966;
  margin: 10px;
  color:#ffffff;
}
```

（6）在 opinion.js 业务逻辑处理文件里，提交意见反馈到后端服务器接口，示例代码如下。

```
var app = getApp();
var host = app.globalData.host;
Page({
  data: {

  },
  onLoad: function (options) {

  },
  formSubmit: function (e) {
    var content = e.detail.value.content;
    if (content == null || content==""){
      wx.showModal({
        title: '提示',
        content: '请填写您的意见或建议',
        showCancel: false
      });
      return;
    }
    var userId = wx.getStorageSync("userId");
    if (userId == null || userId == "") {
      wx.navigateTo({
```

```
          url: '../login/login',
        })
      } else {
        wx.request({
          url: host + '/api/user/saveOpinion',
          method: 'GET',
          data: {
            userId: userId,
            content: content
          },
          header: {
            'Content-Type': 'application/json'
          },
          success: function (res) {
            var code = res.data.code;
            if (code == '0000') {
              wx.showToast({
                title: '保存成功',
                icon: 'success',
                duration: 1000,
                success: function (res) {
                  wx.reLaunch({
                    url: '../me/me'
                  })
                }
              })
            }
          }
        })
      }
    }
  })
})
```

6.10 项目实战：任务 7——实现清除缓存功能

1. 任务目标

通过实现清除缓存功能，学会清理小程序本地缓存数据。

2. 任务实施

在 me.js 业务逻辑处理文件里，添加清理缓存函数，通过在"我的"界面中单击清理缓存菜单可以清理缓存，示例代码如下。

项目实战：实现清除缓存功能

```
var app = getApp();
var host = app.globalData.host;
Page({
  data: {
    nickName: '立即登录'
  },
  onLoad: function(options) {
    this.checkLogin();//校验是否登录
    wx.setNavigationBarTitle({//设置标题
      title: '我的'
    })
  },
  checkLogin: function() {//校验是否登录
```

```
    var userId = wx.getStorageSync("userId");
    if (userId == null || userId == "") {
      wx.navigateTo({
        url: '../login/login',
      })
    } else {
      this.setData({
        nickName: wx.getStorageSync("nickName")
      });
    }
  },
  nav: function(e) {//我的订单跳转
    var id = e.currentTarget.id;
    var status = e.currentTarget.dataset.status;
    wx.navigateTo({
      url: '../myOrder/myOrder?id=' + id + '&status=' + status
    })
  },
  updatePwd: function(e) {//修改密码
    wx.navigateTo({
      url: '../updatePwd/updatePwd'
    })
  },
  opinion: function (e) {//意见反馈
    wx.navigateTo({
      url: '../opinion/opinion'
    })
  },
  clearStore:function(e){//清除缓存
    wx.clearStorageSync();
    wx.showToast({
      title: '清除缓存成功',
      icon: 'success',
      duration: 1000
    })
    wx.reLaunch({
      url: '../me/me'
    })
  }
})
```

6.11 小结

　　注册、登录功能是微信小程序非常基本的功能，本章主要围绕莫凡商城注册、登录功能来讲解如何综合应用微信小程序表单组件、界面交互 API、定时器 API、数据缓存 API、登录相关接口 API 等知识来完成注册功能、登录功能、"我的"界面列表式导航功能、修改密码功能、意见反馈功能和清除缓存功能的设计。

第7章
莫凡商城商品详情页设计

莫凡商城商品详情页设计用来实现商品详情页功能、商品加入购物车功能、商品列表功能、商品详情页分享与转发功能，会用到页面间传递数据及媒体相关组件和媒体相关 API 的使用。

7.1 页面间传递数据

在首页图书商品列表页面中查看图书商品的详情，需要将图书商品的 id 传递给详情页，在商品详情页根据图书商品的 id 来获取图书商品的具体信息，那么如何在页面间传递数据呢？

下面我们一起来实现将图书商品 id 传递给详情页的功能。

（1）在 app.json 文件里添加意见反馈界面路径"pages/goodsDetail/goodsDetail"。

（2）在 index.js 首页业务逻辑处理文件里，添加跳转商品详情页绑定函数，并且把商品 id 作为参数携带，示例代码如下。

慕课视频
页面间传递数据

```
var app = getApp();
var host = app.globalData.host;
Page({
  data: {
    indicatorDots: true,
    autoplay: true,
    interval: 5000,
    duration: 1000,
    imgUrls: [
      "/pages/images/haibao/1.jpg",
      "/pages/images/haibao/2.jpg",
      "/pages/images/haibao/3.jpg"
    ],
    hotList:[ ], //热门书籍列表
    spikeList:[ ], //畅销书籍列表
    bestSellerList:[ ], //秒杀时刻列表
    host: host
  },
  onLoad: function (options) {
    var page = this;
    page.getBannerList();
    page.getBookList();
  },
  getBannerList: function () {
    var page = this;
    wx.request({
      url: host + '/api/banner/getBannerList?type=0',
      method: 'GET',
      data: { },
      header: {
```

```javascript
          'Content-Type': 'application/json'
        },
        success: function (res) {
          var code = res.data.code;
          var list = res.data.data;
          if (code == '0000') {
            var code = res.data.code;
            var list = res.data.data;
            if (code == '0000') {
              var imgUrls = new Array();
              for (var i = 0; i < list.length; i++) {
                imgUrls.push(host + "/" + list[i].url);
              }
              page.setData({ imgUrls: imgUrls });
            }
          }
        }
      })
    },
    getBookList: function () {//获取图书列表方法
      var page = this;
      wx.request({
        url: host + '/api/goods/getHomeGoodsList',
        method: 'GET',
        data: {},
        header: {
          'Content-Type': 'application/json'
        },
        success: function (res) {
          var book = res.data.data;
          //将图书列表数据缓存到本地
          wx.setStorage({
            key: 'book',
            data: book,
          })
          //获取缓存到本地图书列表数据
          book = wx.getStorageSync('book');
          console.log(book);
          var hotList = book.rmjs;//热门书籍列表
          var spikeList = book.mssk;//秒杀时刻书籍列表
          var bestSellerList = book.cxsj;//畅销书籍列表
          page.setData({ hotList: hotList });
          page.setData({ spikeList: spikeList });
          page.setData({ bestSellerList: bestSellerList });
        }
      })
    },
    more:function(e){//查看更多
      var id = e.currentTarget.id;
      wx.navigateTo({
        url: '../goods/goods?id='+id,
      })
    },
    seeDetail: function (e) {//查看商品详情
      var goodsId = e.currentTarget.id;
      wx.navigateTo({
        url: '../goodsDetail/goodsDetail?goodsId=' + goodsId,
      })
    },
    searchInput:function(e){//进入搜索页面
      wx.navigateTo({
```

```
      url: '../search/search',
    })
  }
})
```

（3）在 goodsDetail.js 商品详情业务逻辑处理文件里，通过 onLoad 生命周期函数来获取传递过来的参数，示例代码如下。

```
var app = getApp();
var host = app.globalData.host;
Page({
  data: {

  },
  onLoad: function(e) {
    //从参数 e 里面获取上一个页面携带过来的参数
    var goodsId = e.goodsId;
  },
})
```

通过 onLoad: function（e）函数来获取携带过来的参数，携带过来的值都放在参数 e 里面，通过 e.goodsId 或者其他携带过来的值，就可以实现在页面间传递数据了。

7.2 媒体组件及媒体 API 的应用

微信小程序经常需要实现播放音频、播放视频、相机拍照、实时音视频播放、实时音视频录制等功能，我们可以通过 audio 音频组件及 API 实现音频播放功能；通过 video 视频组件及视频 API 实现视频播放功能；通过 camera 相机组件及相机 API 实现相机拍照功能；通过 liver-player 组件实现实时音视频播放功能；通过 live-pusher 组件实现实时音视频录制功能。本节将详细介绍媒体组件及媒体 API 的应用。

7.2.1 audio 音频组件及音频 API

慕课视频

媒体组件及媒体 API 的应用

1. audio 音频组件

audio 音频组件需要有唯一的 id，根据 id 使用 wx.createAudioContext('myAudio') 创建音频播放的环境，从 1.6.0 版本开始，该组件不再维护。建议使用能力更强的 wx.createInnerAudioContext 接口，具体属性如表 7.1 所示。

表 7.1 audio 音频的属性

属性	类型	默认值	说明
id	string		video 组件的唯一标识符
src	string		要播放音频的资源地址
loop	boolean	false	是否循环播放
controls	boolean	true	是否显示默认控件
poster	string		默认控件上音频封面的图片资源地址，如果 controls 的属性值为 false，则设置 poster 无效
name	string	未知音频	默认控件上的音频名字，如果 controls 的属性值为 false，则设置 name 无效
author	string	未知作者	默认控件上的作者名字，如果 controls 的属性值为 false，则设置 author 无效
binderror	eventHandle		当发生错误时触发 error 事件，detail = {errMsg: MediaError.code}，MediaError.code，错误码：1 为获取资源被用户禁止，2 为网络错误，3 为解码错误，4 为不合适资源

续表

属性	类型	默认值	说明
bindplay	eventHandle		当开始/继续播放时触发 play 事件
bindpause	eventHandle		当暂停播放时触发 pause 事件
bindtimeupdate	eventHandle		当播放进度改变时触发 timeupdate 事件，detail = {currentTime, duration}
bindended	eventHandle		当播放到末尾时触发 ended 事件

示例代码如下。

```
<!-- audio.wxml -->
<audio poster="{{poster}}" name="{{name}}" author="{{author}}" src="{{src}}" id="myAudio" controls loop></audio>

<button type="primary" bindtap="audioPlay">播放</button>
<button type="primary" bindtap="audioPause">暂停</button>
<button type="primary" bindtap="audio14">设置当前播放时间为 14 秒</button>
<button type="primary" bindtap="audioStart">回到开头</button>
```

```
// audio.js
Page({
  onReady: function (e) {
    // 使用 wx.createAudioContext 获取 audio 上下文 context
    this.audioCtx = wx.createAudioContext('myAudio')
  },
  data: {
    poster: 'http://y.gtimg.cn/music/photo_new/T002R300x300M000003rsKF44GyaSk.jpg?max_age=2592000',
    name: '此时此刻',
    author: '许巍',
    src: 'http://ws.stream.qqmusic.qq.com/M500001VfvsJ21xFqb.mp3?guid=ffffffff82def4af4b12b3cd9337d5e7&uin=346897220&vkey=6292F51E1E384E06DCBDC9AB7C49FD713D632D313AC4858BACB8DDD29067D3C601481D36E62053BF8DFEAF74C0A5CCFADD6471160CAF3E6A&fromtag=46',
  },
  audioPlay: function () {//绑定的播放事件
    this.audioCtx.play()
  },
  audioPause: function () {//绑定的暂停事件
    this.audioCtx.pause()
  },
  audio14: function () {//指定多少秒开始播放
    this.audioCtx.seek(14)
  },
  audioStart: function () {//从头播放
    this.audioCtx.seek(0)
  }
})
```

2. AudioContext 音频 API

从基础库 1.6.0 开始，AudioContext 接口停止维护，可以使用 wx.createAudioContext(string id, Object this) 创建 audio 上下文 AudioContext 对象。AudioContext 对象提供以下方法。

（1）AudioContext.pause()暂停音频。

（2）AudioContext.play()播放音频。

（3）AudioContext.seek（number position）跳转到指定位置，position 的单位为秒（s）。

（4）AudioContext.setSrc（string src）设置音频地址。

3. InnerAudioContext 音频 API

使用 wx.createInnerAudioContext()可以创建内部 audio 上下文 InnerAudioContext 对象。

AudioContext 对象提供以下方法。

（1）InnerAudioContext.play()播放音频。

（2）InnerAudioContext.pause()暂停音频。暂停后的音频再播放会从暂停处开始播放。

（3）InnerAudioContext.stop()停止音频。停止后的音频再播放会从头开始播放。

（4）InnerAudioContext.seek（number position）跳转到指定位置。

（5）InnerAudioContext.destroy()销毁当前实例。

（6）InnerAudioContext.onCanplay（function callback）监听音频进入可以播放状态的事件，但不保证后面可以流畅播放。

（7）InnerAudioContext.offCanplay（function callback）取消监听音频进入可以播放状态的事件。

（8）InnerAudioContext.onPlay（function callback）监听音频播放事件。

（9）InnerAudioContext.offPlay（function callback）取消监听音频播放事件。

（10）InnerAudioContext.onPause（function callback）监听音频暂停事件。

（11）InnerAudioContext.offPause（function callback）取消监听音频暂停事件。

（12）InnerAudioContext.onStop（function callback）监听音频停止事件。

（13）InnerAudioContext.offStop（function callback）取消监听音频停止事件。

（14）InnerAudioContext.onEnded（function callback）监听音频自然播放至结束的事件。

（15）InnerAudioContext.offEnded（function callback）取消监听音频自然播放至结束的事件。

（16）InnerAudioContext.onTimeUpdate（function callback）监听音频播放进度更新事件。

（17）InnerAudioContext.offTimeUpdate（function callback）取消监听音频播放进度更新事件。

（18）InnerAudioContext.onError（function callback）监听音频播放错误事件。

（19）InnerAudioContext.offError（function callback）取消监听音频播放错误事件。

（20）InnerAudioContext.onWaiting（function callback）监听音频加载中事件。当音频因为数据不足，需要停下来加载时会触发。

（21）InnerAudioContext.offWaiting（function callback）取消监听音频加载中事件。

（22）InnerAudioContext.onSeeking（function callback）监听音频进行跳转操作的事件。

（23）InnerAudioContext.offSeeking（function callback）取消监听音频进行跳转操作的事件。

（24）InnerAudioContext.onSeeked（function callback）监听音频完成跳转操作的事件。

（25）InnerAudioContext.offSeeked（function callback）取消监听音频完成跳转操作的事件。

4. BackgroundAudioManager 背景音频 API

BackgroundAudioManager 对象实例，可通过 wx.getBackgroundAudioManager 获取。BackgroundAudioManager 提供以下方法。

（1）BackgroundAudioManager.play()播放音频。

（2）BackgroundAudioManager.pause()暂停音频。

（3）BackgroundAudioManager.seek（number currentTime）跳转到指定位置。

（4）BackgroundAudioManager.stop()停止音频。

（5）BackgroundAudioManager.onCanplay（function callback）监听背景音频进入可播放状态事件，但不保证后面可以流畅播放。

（6）BackgroundAudioManager.onWaiting（function callback）监听音频加载中事件。当音频因为数据不足，需要停下来加载时会触发。

（7）BackgroundAudioManager.onError（function callback）监听背景音频播放错误事件。

（8）BackgroundAudioManager.onPlay（function callback）监听背景音频播放事件。

（9）BackgroundAudioManager.onPause（function callback）监听背景音频暂停事件。

（10）BackgroundAudioManager.onSeeking（function callback）监听背景音频开始跳转操作事件。

（11）BackgroundAudioManager.onSeeked（function callback）监听背景音频完成跳转操作事件。

（12）BackgroundAudioManager.onEnded（function callback）监听背景音频自然播放结束事件。

（13）BackgroundAudioManager.onStop（function callback）监听背景音频停止事件。

（14）BackgroundAudioManager.onTimeUpdate（function callback）监听背景音频播放进度更新事件，只有小程序在前台时会回调。

（15）BackgroundAudioManager.onNext（function callback）监听用户在系统音乐播放面板中单击下一曲事件（仅用于 iOS）。

（16）BackgroundAudioManager.onPrev（function callback）监听用户在系统音乐播放面板中单击上一曲事件（仅用于 iOS）。

7.2.2 video 视频组件及视频 API

1. video 视频组件

video 视频组件是用来播放视频的组件，可以控制是否显示默认播放控件（"播放/暂停"按钮、播放进度、时间），可以发送弹幕信息等，video 组件的默认宽度为 300 px，高度为 225 px，宽、高需要通过用 WXSS 设置 width 和 height 来调整，具体属性如表 7.2 所示。

表 7.2 video 视频的属性

属性	类型	默认值	说明
src	string		要播放视频的资源地址
controls	boolean	true	是否显示默认播放控件（"播放/暂停"按钮、播放进度、时间）
danmu-list	Object	Array	弹幕列表
danmu-btn	boolean	false	是否显示弹幕按钮，只在初始化时有效，不能动态变更
enable-danmu	boolean	false	是否展示弹幕，只在初始化时有效，不能动态变更
autoplay	boolean	false	是否自动播放
bindplay	eventHandle		当开始/继续播放时触发 play 事件
bindpause	eventHandle		当暂停播放时触发 pause 事件
bindended	eventHandle		当播放到末尾时触发 ended 事件
bindtimeupdate	eventHandle		播放进度变化时触发，event.detail = {currentTime: '当前播放时间'}。触发频率应该在 250 ms/次
objectFit	string	contain	当视频大小与 video 容器大小不一致时，视频的表现形式：contain 为包含，fill 为填充，cover 为覆盖

示例代码如下。

```
<view class="section tc">
  <video id="myVideo" src="http://wxsnsdy.tc.qq.com/105/20210/snsdyvideodownload?filekey=30280201010421301f0201690402534804102ca905ce620b1241b726bc41dcff44e00204012882540400&bizid=1023&hy=SH&fileparam=302c020101042530230204136ffd93020457e3c4ff02024ef202031e8d7f02030f42400204045a320a0201000400"
  danmu-list="{{danmuList}}" enable-danmu danmu-btn controls></video>
  <view class="btn-area">
    <button bindtap="bindButtonTap">获取视频</button>
    <input bindblur="bindInputBlur"/>
```

```html
    <button bindtap="bindSendDanmu">发送弹幕</button>
  </view>
</view>
```

```javascript
function getRandomColor () {
  let rgb = [ ]
  for (let i = 0 ; i < 3; ++i){
    let color = Math.floor(Math.random() * 256).tostring(16)
    color = color.length == 1 ? '0' + color : color
    rgb.push(color)
  }
  return '#' + rgb.join('')
}

Page({
  onReady: function (res) {
    this.videoContext = wx.createVideoContext('myVideo')
  },
  inputValue: '',
    data: {
        src: '',
    danmuList: [
      {
        text: '第 1s 出现的弹幕',
        color: '#ff0000',
        time: 1
      },
      {
        text: '第 3s 出现的弹幕',
        color: '#ff00ff',
        time: 3
      }]
    },
  bindInputBlur: function(e) {
    this.inputValue = e.detail.value
  },
  bindButtonTap: function() {
    var that = this
    wx.chooseVideo({
      sourceType: ['album', 'camera'],
      maxDuration: 60,
      camera: ['front', 'back'],
      success: function(res) {
        that.setData({
          src: res.tempFilePath
        })
      }
    })
  },
  bindSendDanmu: function () {
    this.videoContext.sendDanmu({
      text: this.inputValue,
      color: getRandomColor()
    })
  }
})
```

界面效果如图 7.1 所示。

图 7.1 视频播放界面

2. video 视频 API

可以使用 wx.createVideoContext（string id, Object this）创建 video 上下文 VideoContext 对象。VideoContext 对象提供以下方法。

（1）VideoContext.play()播放视频。
（2）VideoContext.pause()暂停视频。
（3）VideoContext.stop()停止视频。
（4）VideoContext.seek（number position）跳转到指定位置。
（5）VideoContext.sendDanmu（Object data）发送弹幕。
（6）VideoContext.playbackRate（number rate）设置倍速播放。
（7）VideoContext.requestFullScreen（Object object）进入全屏。
（8）VideoContext.exitFullScreen()退出全屏。
（9）VideoContext.showStatusBar()显示状态栏，仅在 iOS 全屏下有效。
（10）VideoContext.hideStatusBar()隐藏状态栏，仅在 iOS 全屏下有效。
（11）wx.chooseVideo（Object object）拍摄视频或从手机相册中选视频。
（12）wx.saveVideoToPhotosAlbum（Object object） 保存视频到系统相册，支持 MP4 视频格式，调用前需要用户授权 scope.writePhotosAlbum。

7.2.3 camera 相机组件及相机 API

1. camera 相机组件

camera 是相机组件，在使用的时候需要用户授权 scope.camera。camera 相机组件是由客户端创建的原生组件，它的层级是最高的，不能通过 z-index 控制层级，可使用 cover-view、cover-image 覆盖在上面，同一页面只能插入一个 camera 组件，不能在 scroll-view、swiper、picker-view、movable-view 中使用 camera 组件。camera 相机组件的属性如表 7.3 所示。

表 7.3 camera 相机组件的属性

属性	类型	默认值	说明
mode	string	normal	应用模式，只在初始化时有效，不能动态变更，normal 为相机模式、scanCode 为扫码模式

续表

属性	类型	默认值	说明
device-position	string	Back	前置或后置，值为 front、back
flash	string	auto	闪光灯，auto 为自动、on 为打开、off 为关闭、torch 为常亮
frame-size	string	medium	指定期望的相机帧数据尺寸，small 为小尺寸、medium 为中尺寸、large 为大尺寸
bindinitdone	eventhandle		用户不允许使用摄像头时触发
bindscancode	eventhandle		在扫码识别成功时触发，仅在 mode="scanCode"时生效
bindstop	eventhandle		摄像头在非正常终止时触发，如退出后台等情况
binderror	eventhandle		用户不允许使用摄像头时触发

示例代码如下。

```
<camera device-position="back" flash="off" binderror="error" style="width: 100%; height: 300px;"></camera>
<button type="primary" bindtap="takePhoto">拍照</button>
<view>预览</view>
<image mode="widthFix" src="{{src}}"></image>
```

```
Page({
  takePhoto() {
    const ctx = wx.createCameraContext()
    ctx.takePhoto({
      quality: 'high',
      success: (res) => {
        this.setData({
          src: res.tempImagePath
        })
      }
    })
  },
  error(e) {
    console.log(e.detail)
  }
})
```

2. camera 相机 API

可以使用 wx.createCameraContext() 创建 camera 上下文 CameraContext 对象，CameraContext 与页面内唯一的 camera 组件绑定，操作对应的 camera 组件。CameraContext 对象提供以下方法。

（1）CameraContext.onCameraFrame（onCameraFrameCallback callback）获取 Camera 实时帧数据。

（2）CameraContext.takePhoto（Object object）拍摄照片。

（3）CameraContext.startRecord（Object object）开始录像。

（4）CameraContext.stopRecord()结束录像。

（5）CameraFrameListener 是 CameraContext.onCameraFrame()返回的监听器，CameraFrameListener.start()开始监听帧数据，CameraFrameListener.stop()停止监听帧数据。

7.2.4 live-player 实时音视频播放

live-player 为实时音视频播放组件，它的使用是针对特定类目开放的，需要先通过类目审核，再在小程序管理后台中通过"设置"→"接口设置"命令自助开通该组件的权限。目前支持的类目有社交（直播）、教育（在线教育）、医疗（互联网医院、公立医院）、金融[银行、信托、基金、证券/期货、证券/期货投资咨询、保险、征信业务、新三板信息服务平台、股票信息服务平台（港股/美股）、消费金融]、汽车（汽车预售服务）、政府主体账号、工具（视频客服）。live-player 实时音视频播放组件的属性如表 7.4 所示。

表 7.4　live-player 实时音视频播放组件的属性

属性	类型	默认值	说明
src	string		音视频地址，目前仅支持 flv、rtmp 格式
mode	string	live	live（直播）、RTC（实时通话）
autoplay	boolean	false	自动播放
muted	boolean	false	是否静音
orientation	string	vertical	画面方向，可选值有 vertical、horizontal
object-fit	string	contain	填充模式，可选值有 contain、fillCrop
background-mute	Boolean	false	进入后台时是否静音（已废弃，默认退台静音）
min-cache	number	1	最小缓冲区，单位为秒
max-cache	number	3	最大缓冲区，单位为秒
sound-mode	string	speaker	声音输出方式，speaker 为扬声器、ear 为听筒
auto-pause-if-navigate	boolean	true	当跳转到其他小程序页面时，是否自动暂停本页面的实时音视频播放
auto-pause-if-open-native	boolean	true	当跳转到其他微信原生页面时，是否自动暂停本页面的实时音视频播放
bindstatechange	eventHandle		播放状态变化事件，detail = {code}
bindfullscreenchange	eventHandle		全屏变化事件，detail = {direction, fullScreen}
bindnetstatus	eventHandle		网络状态通知，detail = {info}

示例代码如下。

```
<live-player src="https://domain/pull_stream" mode="RTC" autoplay bindstatechange="statechange" binderror="error" style="width: 300px; height: 225px;" />

Page({
  statechange(e) {
    console.log('live-player code:', e.detail.code)
  },
  error(e) {
    console.error('live-player error:', e.detail.errMsg)
  }
})
```

7.2.5　live-pusher 实时音视频录制

live-pusher 为实时音视频录制组件，它的使用需要取得用户授权 scope.camera、scope.record，针对特定类目开放的，需要先通过类目审核，再在小程序管理后台中通过"设置"→"接口设置"命令自助开通该组件的权限。现在支持的类目有社交（直播）、教育（在线视频课程）、医疗（互联网医院、公立医院）、金融[银行、信托、基金、证券/期货、证券/期货投资咨询、保险、征信业务、新三板信息服务平台、股票信息服务平台（港股/美股）、消费金融（金融产品视频客服理赔、金融产品推广直播等）]、汽车（汽车预售服务）、政府主体账号、工具（视频客服）。live-pusher 实时音视频录制组件的属性如表 7.5 所示。

表 7.5　live-pusher 实时视频录制组件的属性

属性	类型	默认值	说明
url	string		推流地址，目前仅支持 flv、rtmp 格式
mode	string	RTC	类型，包括 SD（标清）、HD（高清）、FHD（超清）、RTC（实时通话）

续表

属性	类型	默认值	说明
autopush	boolean	false	自动推流
muted	boolean	false	是否静音
enable-camera	boolean	true	开启摄像头
auto-focus	boolean	true	自动聚集
orientation	string	vertical	垂直 vertical、水平 horizontal
beauty	number	0	美颜
whiteness	number	0	美白，取值范围为 0~9，0 表示关闭
aspect	string	9∶16	宽高比，可选值有 3∶4、9∶16
min-bitrate	number	200	最小码率
max-bitrate	number	1000	最大码率
audio-quality	string	high	高音质（48 kHz）或低音质（16 kHz），值为 high、low
waiting-image	string		进入后台时推流的等待画面
waiting-image-hash	string		等待画面资源的 MD5 值
zoom	boolean	false	调整焦距
device-position	string	front	前置或后置，值为 front、back
background-mute	boolean	false	进入后台时是否静音
mirror	boolean	false	设置推流画面是否镜像
background-mute	boolean	false	进入后台时是否静音
bindstatechange	eventhandle		状态变化事件，detail = {code}
bindnetstatus	eventhandle		网络状态通知，detail = {info}
binderror	eventhandle		渲染错误事件，detail = {errMsg, errCode}
bindbgmstart	eventhandle		背景音开始播放时触发
bindbgmprogress	eventhandle		背景音进度变化时触发，detail = {progress, duration}
bindbgmcomplete	eventhandle		背景音播放完成时触发

示例代码如下。

```
<live-pusher url="https://domain/push_stream" mode="RTC" autopush bindstatechange="statechange" style="width: 300px; height: 225px;" />

Page({
  statechange(e) {
    console.log('live-pusher code:', e.detail.code)
  }
})
```

7.3 项目实战：任务 13——实现商品详情页功能

慕课视频

项目实战：实现商品详情页功能

1. 任务目标

通过实现商品详情页功能，来巩固海报轮播效果设计、video 视频组件的使用、页面布局设计、页签切换效果设计、动态获取数据、动态数据绑定这些常用知识点。

莫凡商城商品详情页的功能可以分解为海报轮播效果布局设计、使用 video 视频组件介绍商品、图书详情布局设计、图书详情与出版信息页签切换、动态获取图书详情页数据、动态数据绑定详情页，如图 7.2 和图 7.3 所示。

图 7.2　商品详情页 1

图 7.3　商品详情页 2

2. 任务实施

（1）在 goodsDetail.wxml 文件的商品详情页里进行商品详情布局设计，具体代码如下。

```
<view class="content">
  <view class="haibao">
    <swiper indicator-dots="{{indicatorDots}}" autoplay="{{autoplay}}" interval="{{interval}}" duration="{{duration}}" class="swiperHeight">
      <block wx:for="{{goodsDetail.roundPlayPicList}}">
        <swiper-item>
          <image src="{{item}}" class="silde-image" mode="aspectFill"></image>
        </swiper-item>
      </block>
    </swiper>
  </view>
  <view class="title"><text class="tip">莫凡自营</text>{{goodsDetail.goodsName}}</view>
  <view class="desc">{{goodsDetail.briefIntroduction}}</view>
  <view class="price"><text class="symbol">￥</text><text class="account">{{goodsDetail.goodsPrice}}</text> <text>定价:</text><text class="oldPrice">￥{{goodsDetail.goodsCost}}</text></view>
  <view class="hr"></view>
  <view class="items">
    <view class="item">
      <view class="term">作者</view>
      <view>{{goodsDetail.author}}</view>
    </view>
    <view class="line"></view>
    <view class="item">
      <view class="term">出版</view>
      <view>人民邮电出版社, {{goodsDetail.publishTime}}</view>
    </view>
  </view>
  <view class="hr"></view>
  <view class="mark">
    <view><image src="/pages/images/icon/support.png" style="width:15px;height:15px;"></image>
    <text class="searchContent">正品保障</text></view>
    <view><image src="/pages/images/icon/support.png" style="width:15px;height:15px;"></image>
    <text class="searchContent">支持礼品卡</text></view>
```

```
        <view><image src="/pages/images/icon/support.png" style="width:15px;height:15px;"></image>
        <text class="searchContent">支持 7 日无理由退货</text></view>
        <view><image src="/pages/images/icon/support.png" style="width:15px;height:15px;"></image>
        <text class="searchContent">礼品包装</text></view>
</view>
<view class="line"></view>
<view class="items">
    <view class="item">
        <view class="term">莫凡配送</view>
        <view class="nav">运费 8 元, 满 66 元包邮</view>
    </view>
</view>
<view class="hr"></view>
<view class="items">
    <view class="item">
    <view class="term">数量</view>
        <view class="priceInfo">
            <view class="minus" id="{{goodsDetail.id}}" bindtap="minusGoods">-</view>
            <view class="count">{{num}}</view>
            <view class="add" id="{{goodsDetail.id}}" bindtap="addGoods">+</view>
        </view>
    </view>
</view>
<view class="hr"></view>
<view class="tab">
    <view class="{{currentTab==0?'select':'normal'}}" id="0" bindtap="switchNav">图书详情</view>
    <view class="{{currentTab==1?'select':'normal'}}" id="1" bindtap="switchNav">出版信息</view>
</view>
<view>
    <swiper current="{{currentTab}}" style="height:2000px;">
        <swiper-item>
            <view class="detail">
                <block wx:for="{{goodsDetail.infoPicList}}">
                    <image src="{{item}}" mode="widthFix"></image>
                </block>
            </view>
        </swiper-item>
        <swiper-item>
            <view class="items">
                <view class="item">
                    <view class="term">书名</view>
                    <view>{{goodsDetail.bookName}}</view>
                </view>
                <view class="line"></view>
                <view class="item">
                    <view class="term">ISBN</view>
                    <view>{{goodsDetail.isbn}}</view>
                </view>
                <view class="line"></view>
                <view class="item">
                    <view class="term">作者</view>
                    <view>{{goodsDetail.author}}</view>
                </view>
                <view class="line"></view>
                <view class="item">
                    <view class="term">出版社</view>
                    <view>{{goodsDetail.bookConcern}}</view>
                </view>
                <view class="line"></view>
                <view class="item">
```

```xml
          <view class="term">出版时间</view>
          <view>{{goodsDetail.publishTime}}</view>
        </view>
        <view class="line"></view>
        <view class="item">
          <view class="term">版次</view>
          <view>{{goodsDetail.edition}}</view>
        </view>
        <view class="line"></view>
        <view class="item">
          <view class="term">开本</view>
          <view>{{goodsDetail.paperSize}}</view>
        </view>
        <view class="line"></view>
        <view class="item">
          <view class="term">纸张</view>
          <view>{{goodsDetail.paper}}</view>
        </view>
        <view class="line"></view>
        <view class="item">
          <view class="term">包装</view>
          <view>{{goodsDetail.packing}}</view>
        </view>
        <view class="line"></view>
        <view class="item">
          <view class="term">是否套装</view>
          <view wx:if="{{goodsDetail.isSuit==0}}">是</view>
          <view wx:else>否</view>
        </view>
        <view class="line"></view>
      </view>
    </swiper-item>
  </swiper>
</view>
<view class="hr"></view>
<view class="bottom">
  <view class="cart" bindtap='seeCart'><image src="/pages/images/icon/cart.png"></image>
  <text class="label" wx:if="{{cartNum > 0}}">{{cartNum}}</text></view>
  <view class="intocart" bindtap='intocart' id="{{goodsDetail.id}}">加入购物车</view>
  <view class="buy" bindtap="buy" id="{{goodsDetail.id}}">立即购买</view>
</view>

</view>
```

（2）在 goodsDetail.wxss 样式文件里进行商品详情页样式渲染，具体代码如下。

```css
.content{
  width: 100%;
  font-family: "Microsoft YaHei";
}
.haibao{
    text-align: center;
    width: 100%;
}
.swiperHeight{
  height: 250px;
}
.silde-image{
    width: 100%;
    height: 250px;
}
.title{
```

```css
    padding: 10px;
    font-size: 15px;
    font-weight: bold;
    height: 20px;
    line-height: 20px;
}
.tip{
    font-size: 11px;
    padding:3px;
    background-color: #009966;
    color:#ffffff;
    font-weight: normal;
    border-radius: 15px;
    margin-right: 10px;
}
.desc{
    padding-left: 10px;
    padding-right: 10px;
    font-size: 12px;
    color: #999999;
}
.price{
    padding: 10px;
    font-size: 12px;
    color: #999999;
}
.symbol{
    color: red;
    font-size: 14px;
    font-weight: bold;
}
.account{
    color: red;
    font-size: 18px;
    font-weight: bold;
    margin-left: 2px;
    margin-right: 20px;
}
.oldPrice{
    text-decoration: line-through;
}
.items{
    padding-left:10px;
    padding-right:10px;
}
.item{
    display: flex;
    flex-direction: row;
    height: 40px;
    line-height: 40px;
    font-size: 13px;
}
.term{
    width: 70px;
}
.priceInfo{
    display: flex;
    flex-direction: row;
    align-items: center;
}
```

```css
.minus, .add{
    border: 1px solid #cccccc;
    width: 25px;
    height: 20px;
    line-height: 20px;
    color: #009966;
    text-align: center;
    font-weight: bold;
    font-size: 15px;
}
.count{
    width: 30px;
    height: 20px;
    line-height: 20px;
    text-align: center;
    border-top: 1px solid #cccccc;
    border-bottom: 1px solid #cccccc;
}
.hr{
    height: 10px;
    background-color: #dddddd;
}
.line{
    height: 1px;
    width: 100%;
    background-color: #dddddd;
    opacity: 0.2;
}
.mark{
   font-size: 12px;
   padding: 10px;
   height: 20px;
   line-height: 20px;
   display: flex;
   flex-direction: row;
}
.mark view{
    margin-right: 5px;
}
.nav{
    font-size: 12px;
    color: #666666;
}
.tab{
   display: flex;
   flex-direction: row;
   font-size: 13px;
   border-bottom:1px solid #f2f2f2;
}
.select{
    color:#009966;
    display: inline-block;
    line-height: 80rpx;
    width: 50%;
    text-align: center;
    border-bottom: 5rpx solid #009966;
    font-weight: bold;
}
.normal{
    display: inline-block;
    line-height: 80rpx;
```

```css
      width: 50%;
      text-align: center;
}
.detail{
   width: 90%;
   margin: 0 auto;
   margin-top:10px;
}
.detail image{
   width: 100%;
}
.bottom{
   background-color:#ffffff;
   height: 50px;
   position: fixed;
   bottom: 0px;
   width: 100%;
   display: flex;
   flex-direction: row;
}
.cart{
   width: 20%;
   height: 100%;
   text-align: center;
   line-height: 80px;
}
.cart image{
   width: 40px;
   height: 40px;
}
.intocart{
    width: 40%;
    background-color: #ffcc00;
    color: #ffffff;
    font-size:16px;
    text-align: center;
    line-height: 50px;
}
.buy{
   width: 40%;
   background-color: #009966;
   color: #ffffff;
   font-size:16px;
   text-align: center;
   line-height: 50px;
}
.label{
   position: absolute;
   border: 1px solid red;
   font-size: 10px;
   color: red;
   height: 12px;
   line-height:12px;
   width: 20px;
   text-align: center;
   border-radius: 8px;
   left:38px;
   top:5px;
   background-color: #ffffff;
}
```

（3）在 goodsDetail.js 业务逻辑处理文件中进行商品详情页数据获取，具体代码如下。

```js
var app = getApp();
var host = app.globalData.host;
Page({
   data: {
      indicatorDots: true,
      autoplay: true,
```

```javascript
    interval: 5000,
    duration: 1000,
    imgUrls: [
      "/pages/images/books/hot-1.jpg"
    ],
    currentTab: 0,
    goodsDetail: null,
    num: 1,
    cartNum: 0
  },
  onLoad: function(e) {
    var goodsId = e.goodsId;
    this.loadGoodsDetail(goodsId);
  },
  loadGoodsDetail: function(goodsId) {//获取商品详情
    if (goodsId != "") {
      var that = this;
      wx.request({
        url: host + '/api/goods/getGoodsDetail',
        method: 'GET',
        data: {
          "goodsId": goodsId
        },
        header: {
          'Content-Type': 'application/json'
        },
        success: function(res) {
          var goodsDetail = res.data.data;
          that.setData({
            goodsDetail: goodsDetail
          });
        }
      })
    }
  },
  switchNav: function(e) {//图书详情和出版信息页签切换
    var index = e.currentTarget.id;
    this.setData({
      currentTab: index
    });
  },
  buy: function(e) {//立即购买页面跳转
    var goodsId = e.currentTarget.id;
    var userId = wx.getStorageSync("userId");
    if(userId != ''){
      wx.navigateTo({
        url: '../buy/buy?goodsId=' + goodsId + '&num=' + this.data.num
      })
    }else{
      wx.navigateTo({
        url: '../login/login',
      })
    }
  },
  addGoods: function(e) {//添加商品数量
    var num = this.data.num;
    this.setData({
      num: num + 1
    });
  },
  minusGoods: function(e) {//减少商品数量
    var num = this.data.num;
    if (num > 1) {
      this.setData({
        num: num - 1
      });
    }
  },
})
```

7.4 项目实战：任务14——实现商品加入购物车功能

1. 任务目标

通过实现商品加入购物车功能，来巩固导航跳转、购物车页面布局、动态获取数据、动态商品加入购物车等知识。

在莫凡商城商品详情页中，我们可以将商品加入到购物车里，这也是商城常用的设计方式，如图 7.4 所示。

图 7.4　加入购物车

2. 任务实施

（1）在 app.json 文件里添加购物车页面路径"pages/shoppingcart/shoppingcart"。

（2）在 goodsDetail.js 业务逻辑处理文件里添加跳转到购物车页面函数、商品加入购物车函数、查看购物车商品列表函数，具体代码如下。

```
var app = getApp();
var host = app.globalData.host;
Page({
  data: {
    indicatorDots: true,
    autoplay: true,
    interval: 5000,
    duration: 1000,
    imgUrls: [
      "/pages/images/books/hot-1.jpg"
    ],
    currentTab: 0,
    goodsDetail: null,
    num: 1,
    cartNum: 0
```

```js
  },
  onLoad: function(e) {
    var goodsId = e.goodsId;
    this.loadGoodsDetail(goodsId);
    this.loadCart();
  },
  loadGoodsDetail: function(goodsId) {//获取商品详情
    if (goodsId != "") {
      var that = this;
      wx.request({
        url: host + '/api/goods/getGoodsDetail',
        method: 'GET',
        data: {
          "goodsId": goodsId
        },
        header: {
          'Content-Type': 'application/json'
        },
        success: function(res) {
          var goodsDetail = res.data.data;
          that.setData({
            goodsDetail: goodsDetail
          });
        }
      })
    }
  },
  switchNav: function(e) {//图书详情和出版信息页签切换
    var index = e.currentTarget.id;
    this.setData({
      currentTab: index
    });
  },
  buy: function(e) {//立即购买页面跳转
    var goodsId = e.currentTarget.id;
    var userId = wx.getStorageSync("userId");
    if(userId != ''){
      wx.navigateTo({
        url: '../buy/buy?goodsId=' + goodsId + '&num=' + this.data.num
      })
    }else{
      wx.navigateTo({
        url: '../login/login',
      })
    }
  },
  addGoods: function(e) {//添加商品数量
    var num = this.data.num;
    this.setData({
      num: num + 1
    });
  },
  minusGoods: function(e) {//减少商品数量
    var num = this.data.num;
    if (num > 1) {
      this.setData({
        num: num - 1
      });
    }
  },
  intocart: function(e) {//加入购物车
    var that = this;
```

```javascript
      var goodsId = e.currentTarget.id;
      var userId = wx.getStorageSync("userId");
      if (userId != "") {
        wx.request({
          url: host + '/api/cart/saveShoppingCart',
          method: 'GET',
          data: {
            'userId': userId,
            'goodsId': goodsId,
            'type': '0'
          },
          header: {
            'Content-Type': 'application/json'
          },
          success: function(res) {
            var code = res.data.code;
            if (code == '0000') {
              that.loadCart();
            }
          }
        })
      } else {
        wx.redirectTo({
          url: '../login/login'
        })
      }
    },
    seeCart: function (e) {//查看购物车
      console.log(e)
      wx.redirectTo({
        url: '../shoppingcart/shoppingcart'
      })
    },
    loadCart: function() {//获取购物车列表
      var that = this;
      var userId = wx.getStorageSync("userId");
      if (userId != "") {
        wx.request({
          url: host + '/api/cart/getShoppingCartList',
          method: 'GET',
          data: {
            'userId': userId,
            'type': '0'
          },
          header: {
            'Content-Type': 'application/json'
          },
          success: function(res) {
            console.log(res);
            var code = res.data.code;
            if (code == '0000') {
              var ret = res.data.data;
              that.setData({
                cartNum: ret.length
              });
            }
          }
        })
      }else{
        wx.redirectTo({
          url: '../login/login'
        })
      }
    }
  }
})
```

7.5 项目实战：任务 15——实现购物车列表功能

1. 任务目标

通过实现购物车列表功能，来巩固商品列表动态循环渲染及动态添加或减少商品数量等知识。

莫凡商城购物车页面用来显示购物车商品列表，包括商品名称、商品价格、商品数量，如图 7.5 所示。

图 7.5 购物车页面

2. 任务实施

（1）在 shoppingcart.wxml 购物车商品页面文件里进行购物车页面布局设计，具体代码如下。

```
<view class="content">
  <view class="hr"></view>
  <view class="items">
    <radio-group bindchange="radioChange">
      <block wx:for="{{carts}}">
        <view class="item">
          <view class="icon">
            <radio value="{{item.id}}" checked="{{selected}}"/>
          </view>
          <view class="pic">
            <image src="{{item.listPic}}" style="width:70px;height:87px;"></image>
          </view>
          <view class="order">
            <view class="title">{{item.goodsName}}</view>
            <view class="priceInfo">
              <view class="price">¥{{item.goodsPrice}}</view>
              <view class="minus" id="{{item.id}}" bindtap="minusGoods">-</view>
              <view class="count">{{item.num}}</view>
              <view class="add" id="{{item.id}}" bindtap="addGoods">+</view>
            </view>
          </view>
        </view>
        <view class="line"></view>
      </block>
```

慕课视频

项目实战：实现购物车列表功能

```
            </radio-group>
            <view>

            </view>
        </view>

        <view class="bottom">
            <checkbox-group bindchange="checkAll">
                <view class="all">
                    <view class="selectAll">
                        商品总价
                    </view>
                    <view class="total">
                        ￥{{totalPrice}}元
                    </view>
                    <view class="opr" bindtap="buy">
                        去结算
                    </view>
                </view>
            </checkbox-group>
        </view>
</view>
```

（2）在 shoppingcart.wxss 样式文件里对购物车页面进行样式渲染，具体代码如下。

```
.content{
    font-family: "Microsoft YaHei";
    height: 600px;
    background-color: #F9F9F8;
}
.hr{
    height: 12px;
}
.line{
    border: 1px solid #cccccc;
    opacity: 0.2;
}
.items{
    background-color: #ffffff;
}
.item{
    display: flex;
    flex-direction: row;
    padding:10px;
    align-items: center;
}
.order{
    width: 100%;
    height: 87px;
    margin-left: 5px;
}
.title{
    font-size: 15px;
}
.title image{
  width: 15px;
  height: 20px;
  position: absolute;
  right: 10px;
}
.priceInfo{
    display: flex;
    flex-direction: row;
    margin-top:30px;
}
.price{
    width:65%;
    font-size: 15px;
    color: #ff0000;
```

```css
        text-align: left;
}
.minus, .add{
        border: 1px solid #cccccc;
        width: 25px;
        height: 20px;
        line-height: 17px;
        color: #009966;
        text-align: center;
        font-weight: bold;
        font-size: 15px;
}
.count{
        width: 30px;
        height: 20px;
        line-height: 20px;
        text-align: center;
        border-top: 1px solid #cccccc;
        border-bottom: 1px solid #cccccc;
        font-size: 13px;
}
.all{
        display: flex;
        flex-direction: row;
        height: 60px;
        align-items: center;
        padding-left: 10px;
}
.selectAll{
        width: 80px;
        text-align: center;
        font-size: 15px;
        font-weight: bold;
}
.total{
        width: 200px;
        font-size: 15px;
        color: #ff0000;
        font-weight: bold;
}
.opr{
        position: absolute;
        right: 0px;
        width: 120px;
        font-size: 15px;
        font-weight: bold;
        background-color: #009966;
        height: 60px;
        text-align: center;
        line-height: 60px;
        color: #ffffff;
}
.bottom{
    background-color:#ffffff;
    height: 60px;
    position: fixed;
    bottom: 0px;
    width: 100%;
    display: flex;
    flex-direction: row;
}
```

（3）在 shoppingcart.js 业务逻辑处理文件里动态获取购物车里的商品列表，具体代码如下。

```js
var app = getApp();
var host = app.globalData.host;
Page({
    data:{
        carts:[ ],
```

```
    selected:false,
    selectedAll:true,
    totalPrice:0,
    num:1,
    goodsId:''
  },
  onLoad:function(){
    this.loadCarts();
  },
  loadCarts:function(){//加载购物车商品列表
    var page = this;
    var userId = wx.getStorageSync("userId");
    if (userId == null || userId ==""){
      wx.navigateTo({
        url: '../login/login',
      })
    }else{
      wx.request({
        url: host + '/api/cart/getShoppingCartList',
        method: 'GET',
        data: {
          userId: userId,
          type: 0
        },
        header: {
          'Content-Type': 'application/json'
        },
        success: function (res) {
          var carts = res.data.data;
          console.log(carts);
          page.setData({ carts: carts });
        }
      })
    }
  },
  radioChange:function(e){//选择结算商品并计算价格
    console.log(e);
    var id = e.detail.value;
    this.computePrice(id);
  },
  addGoods: function (e) {//添加商品
    var id = e.target.id;
    var carts = this.data.carts;
    for (var i = 0; i < carts.length;i++){
      var cart = carts[i];
      if(id == cart.id){
          cart.num = cart.num+1;
          this.updateCartNum(cart.id, cart.num);
          this.computePrice(id);
          break;
      }
    }
    this.loadCarts();
  },
  minusGoods: function (e) {//减少商品
    var id = e.target.id;
    var carts = this.data.carts;
    for (var i = 0; i < carts.length; i++) {
      var cart = carts[i];
      if (id == cart.id) {
        if (cart.num > 1){
          cart.num = cart.num - 1;
          this.updateCartNum(cart.id, cart.num);
          this.computePrice(id);
          break;
        }
      }
```

```
    }
    this.loadCarts();
},
updateCartNum:function(cartId, num){//更新购物车数量
    wx.request({
      url: host + '/api/cart/updateCartNum',
      method: 'GET',
      data: {
        cartId: cartId,
        num: num
      },
      header: {
        'Content-Type': 'application/json'
      },
      success: function (res) {    }
    })
},
buy:function(){//跳转到结算页面
    var goodsId = this.data.goodsId;
    var userId = wx.getStorageSync("userId");
    if (goodsId == '' || goodsId == null) {
      wx.showModal({
        title: '提示',
        content: '请选择结算商品',
        showCancel: false
      })
    } else {
      wx.navigateTo({
        url: '../buy/buy?goodsId=' + goodsId + '&num=' + this.data.num
      })
    }
},
computePrice: function (id) {//计算商品价格
    //计算商品价格
    var carts = this.data.carts;
    var totalPrice = 0;
    for (var i = 0; i < carts.length; i++) {
      var cart = carts[i];
      if(cart.id==id){
        totalPrice += cart.goodsPrice * cart.num;
        this.setData({ goodsId: cart.goodsId});
        this.setData({ num: cart.num });
        break;
      }
    }
    this.setData({ totalPrice: totalPrice.toFixed(2) });
}
})
```

7.6 商品详情页分享与转发 API 的应用

通常开发者希望转发出去的小程序被二次打开的时候能够使用户获取到一些信息，如群的标识。现在通过调用 wx.showShareMenu 并且设置 withShareTicket 为 true，可以实现当用户将小程序转发到任一群聊之后，此转发卡片在群聊中被其他用户打开时，在 App.onLaunch 或 App.onShow 获取一个 shareTicket。通过调用 wx.getShareInfo 接口传入此 shareTicket 可以获取到转发信息。

微信小程序提供了以下 4 个与转发相关的 API。

（1）wx.showShareMenu（Object object）：显示当前页面的"转发"按钮。

（2）wx.hideShareMenu（Object object）：隐藏"转发"按钮。

（3）wx.updateShareMenu（Object object）：更新转发属性。

（4）wx.getShareInfo（Object object）：获取转发详细信息。

通过 onShareAppMessage（Object object）监听事件来实现转发，监听用户单击页面内"转发"按钮（button 组件 open-type="share"）或右上角菜单中"转发"按钮的行为，并自定义转发内容。OnShareAppMessgae 返回值的属性如表 7.6 所示。

表 7.6 **onShareAppMessage** 返回值的属性

属性	类型	说明
title	转发标题	当前小程序名称
path	转发路径	当前页面 path，必须是以"/"开头的完整路径
imageUrl	自定义图片路径，可以是本地文件路径、代码包文件路径或者网络图片路径；支持 PNG 及 JPG；显示图片长宽比为 5 : 4	使用默认截图

示例代码如下。

```
Page({
  onShareAppMessage: function (res) {
    if (res.from === 'button') {
      // 来自页面内"转发"按钮
      console.log(res.target)
    }
    return {
      title: '自定义转发标题',
      path: '/page/user?id=123'
    }
  }
})
```

下面一起实现将商品详情页分享出去的功能。在 goodsDetail.js 商品业务逻辑处理页面里添加分享监听事件，具体代码如下。

```
onShareAppMessage: function () {//用户单击右上角分享
  var goodsDetail = this.data.goodsDetail;
  return {
    title: goodsDetail.goodsName,   // 分享标题
    path: '/pages/goodsDetail/goodsDetail' // 分享路径
  }
}
```

7.7 小结

本章主要实现了莫凡商城商品详情页设计、商品加入购物车功能、购物车列表功能及商品详情分享功能，应用到了页面间传递数据、海报轮播效果、video 视频组件及 API 的应用、页签切换效果、发起网络请求获取商品详情数据、发起网络请求获取购物车商品列表数据、商品详情页分享等知识。

第8章 莫凡商城获取收货地址功能设计

获取收货地址是 App 和小程序经常会用到的一个功能,莫凡商城也需要从用户那里获取商品的收货地址。获取收货地址会应用到与位置相关的 API、收货地址 API 和地图组件及与地图相关的 API。本章我们综合应用这些知识来完成莫凡商城获取收货地址功能的设计。

8.1 位置 API

慕课视频
位置 API

微信小程序的位置 API 包括查看位置 API、获得位置 API、打开位置 API、开启/停止接收位置信息 API、监听实时地理位置 API,运用这些 API 可以完成和位置相关的设计。

8.1.1 查看位置、获得位置、打开位置

微信小程序提供了 wx.getLocation(OBJECT) 获得当前位置、wx.chooseLocation(OBJECT) 选择位置、wx.openLocation(OBJECT) 打开位置 3 个 API。

1. wx.getLocation(OBJECT) 获得当前位置

使用 wx.getLocation(OBJECT) 可以获得当前位置信息,包括当前位置的地理坐标、移动速度,用户离开小程序后,此接口无法调用。

接口需要传递位置类型 type,默认为 wgs84,返回 GPS 坐标,gcj02 返回可用于 wx.openLocation 的坐标。

接口调用成功后,返回参数说明如表 8.1 所示。

表 8.1 wx.getLocation 返回参数说明

参数	说明
latitude	纬度,浮点数,范围为-90~90,负数表示南纬
longitude	经度,浮点数,范围为-180~180,负数表示西经
speed	速度,浮点数,单位为米/秒
accuracy	位置的精确度
altitude	高度,单位为米
verticalAccuracy	垂直精度,单位为米(Android 无法获取,返回 0)
horizontalAccuracy	水平精度,单位为米

示例代码如下。

```
Page({
  onLoad:function(){
```

```
wx.getLocation({
  type: 'wgs84',
  success: function(res) {
    var latitude = res.latitude;
    console.log("纬度="+latitude);
    var longitude = res.longitude;
    console.log("经度="+longitude);
    var speed = res.speed;
    console.log("速度="+speed);
    var accuracy = res.accuracy;
    console.log("精确度="+accuracy);
  }
})
  }
})
```

2. wx.chooseLocation（OBJECT）选择位置

使用 wx.chooseLocation 打开地图来选择位置，调用前需要用户授权 scope.userLocation，接口调用成功后返回参数说明如表 8.2 所示。

表 8.2　wx.chooseLocation 返回参数说明

参数	说明
latitude	纬度，浮点数，范围为-90～90，负数表示南纬
longitude	经度，浮点数，范围为-180～180，负数表示西经
name	位置名称
address	详细地址

示例代码如下。

```
Page({
  onLoad:function(){
    wx.chooseLocation({
      success: function(res){
        console.log(res);
      }
    })
  }
})
```

3. wx.openLocation（OBJECT）打开位置

使用 wx.openLocation（OBJECT）接口可以打开微信内置地图查看位置，具体参数如表 8.3 所示。

表 8.3　wx.openLocation 返回参数说明

属性	类型	是否必填	说明
latitude	Float	是	纬度，范围为-90～90，负数表示南纬
longitude	Float	是	经度，范围为-180～180，负数表示西经
scale	INT	否	缩放比例，范围为5～18，默认为18
name	string	否	位置名
address	string	否	地址的详细说明
success	Function	是	接口调用成功的回调函数
fail	Function	否	接口调用失败的回调函数
complete	Function	否	接口调用结束的回调函数（调用成功、失败都会执行）

示例代码如下。

```
Page({
  onLoad:function(){
    wx.getLocation({
      type: 'gcj02',    //返回可以用于 wx.openLocation 的经纬度
      success: function(res) {
        var latitude = res.latitude
        var longitude = res.longitude
        wx.openLocation({
          latitude: latitude,
          longitude: longitude,
          scale: 28
        })
      }
    })
  }
})
```

界面效果如图 8.1 所示。

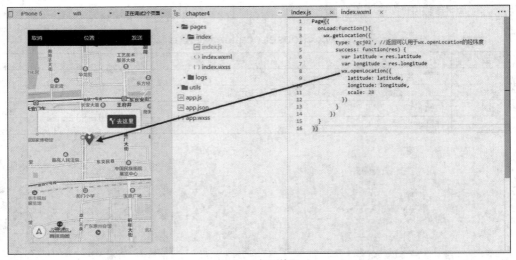

图 8.1 打开位置

8.1.2 开启/停止接收位置信息

微信小程序提供 wx.startLocationUpdate（Object object）来开启小程序进入前台时接收位置信息，调用前需要用户授权 scope.userLocation。

微信小程序提供 wx.startLocationUpdateBackground（Object object）来开启小程序进入前、后台时均接收位置信息，调用前需要用户授权 scope.userLocationBackground，授权以后，小程序在运行中或进入后台时均可接受位置信息变化。

微信小程序提供 wx.stopLocationUpdate（Object object）来关闭实时位置变化的监听，前、后台都停止信息接收。

8.1.3 监听实时地理位置

微信小程序用 wx.onLocationChange（function callback）API 来监听实时地理位置变化事件，需结合 wx.startLocationUpdateBackground、wx.startLocationUpdate 使用。返回参数如表 8.4 所示。

表 8.4　wx.onLocationChange（function callback）返回参数说明

参数	说明
latitude	纬度，浮点数，范围为-90～90，负数表示南纬
longitude	经度，浮点数，范围为-180～180，负数表示西经
speed	速度，单位为米/秒
accuracy	位置的精确度
altitude	高度，单位为米
verticalAccuracy	垂直精度，单位为米（Android 无法获取，返回 0）
horizontalAccuracy	水平精度，单位为米

示例代码如下。

```
const _locationChangeFn = function(res) {
  console.log('location change', res)
}
wx.onLocationChange(_locationChangeFn)
wx.offLocationChange(_locationChangeFn)
```

微信小程序使用 wx.offLocationChange（function callback）API 来取消监听实时地理位置变化事件。

8.2　收货地址 API

慕课视频
收货地址 API

微信小程序提供了收货地址 API，用来获取用户的收货地址，调出用户编辑收货地址的原生界面，并在编辑完成后返回用户选择的地址，调用前需要用户授权 scope.address。收货地址 API 接口调用成功后的返回值如表 8.5 所示。

表 8.5　收货地址返回值

属性	类型	说明
userName	string	收货人姓名
postalCode	string	邮编
provinceName	string	国标收货地址第一级地址
cityName	string	国标收货地址第二级地址
countyName	string	国标收货地址第三级地址
detailInfo	string	详细收货地址信息
nationalCode	string	收货地址国家码
telnumber	string	收货人手机号码
errMsg	string	错误信息

示例代码如下。

```
Page({
  onLoad: function() {
    wx.chooseAddress({
      success(res) {
        console.log(res.userName)
        console.log(res.postalCode)
        console.log(res.provinceName)
        console.log(res.cityName)
        console.log(res.countyName)
        console.log(res.detailInfo)
        console.log(res.nationalCode)
        console.log(res.telnumber)
      }
    })
  }
})
```

8.3 地图组件及地图 API

8.3.1 map 地图组件

微信小程序提供了地图功能，通过 map 地图组件来开发与地图有关的应用，如共享单车、滴滴打车、外卖配送查询等，在地图上可以标记覆盖物并指定一系列的坐标位置，如共享单车应用的地图上会标识共享单车的位置。

map 地图组件的具体属性如表 8.6 所示。

表 8.6 map 地图组件的属性

属性	类型	默认值	说明
longitude	number		中心经度
latitude	number		中心纬度
scale	number	16	缩放级别，取值范围为 3~20
markers	Array		标记点
covers	Array		即将移除，请使用 markers
autoplay	boolean		是否自动播放
polyline	Array		路线
circles	Array		圆
controls	Array		控件（即将废弃，建议使用 cover-view 代替）
include-points	Array		缩放视野以包含所有给定的坐标点
show-location	boolean		显示带有方向的当前定位点
polygons	Array.<polygon>		多边形
subkey	string		个性化地图使用的 key
layer-style	number	1	个性化地图配置的 style，不支持动态修改
rotate	number	0	旋转角度，范围为 0~360，地图正北和设备 y 轴角度的夹角
skew	number	0	倾斜角度，范围为 0~40，关于 z 轴的倾角
enable-3D	boolean	false	展示 3D 楼块（工具暂不支持）
show-compass	boolean	false	显示指南针
show-scale	boolean	false	显示比例尺，工具暂不支持
enable-overlooking	boolean	false	开启俯视
enable-zoom	boolean	true	是否支持缩放
enable-scroll	boolean	true	是否支持拖动
enable-rotate	boolean	false	是否支持旋转
enable-satellite	boolean	false	是否开启卫星图
enable-traffic	boolean	false	是否开启实时路况
setting	object		配置项
bindcallouttap	eventhandle		单击标记点对应的气泡时触发，e.detail = {markerId}
bindmarkertap	eventHandle		单击标记点时触发，e.detail = {markerId}
bindcontroltap	eventHandle		单击控件时触发，e.detail = {controlId}
bindregionchange	eventHandle		视野发生变化时触发
bindtap	eventHandle		单击地图时触发
bindupdated	eventhandle		在地图渲染更新完成时触发
bindpoitap	eventhandle		单击地图 poi 点时触发，e.detail = {name, longitude, latitude}

markers 标记点用于在地图上显示标记的位置，如表 8.7 所示。

表 8.7 marker 地图标记的属性

属性	说明	类型	是否必填	备注
id	标记点 id	number	否	marker 单击事件回调会返回此 id。建议为每个 marker 设置 number 类型 id，保证更新 marker 时有更好的性能
latitude	纬度	number	是	浮点数，范围为-90~90
longitude	经度	number	是	浮点数，范围为-180~180
title	标注点名	string	否	单击时显示，callout 存在时将被忽略
zIndex	显示层级	number		
iconPath	显示的图标	string	是	项目目录下的图片路径，支持相对路径的写法，以 "/" 开头则表示相对小程序根目录；也支持临时路径和网络图片
rotate	旋转角度	number	否	顺时针旋转的角度，范围为 0~360，默认为 0
alpha	标注的透明度	number	否	默认为 1，无透明
width	标注图标的宽度	number	否	默认为图片实际宽度
height	标注图标的高度	number	否	默认为图片实际高度
callout	自定义标记点上方的气泡窗口	Object		
label	为标记点旁边增加标签	Object		
anchor	经、纬度在标注图标的锚点，默认为底边中点	Object		{x, y}，x 表示横向（0~1），y 表示竖向（0~1），{x: .5, y: 1}表示底边中点
aria-label	无障碍访问，（属性）元素的额外描述	string		

polyline 用来指定一系列坐标点，从数组第一项连线至最后一项，如表 8.8 所示。

表 8.8 polyline 坐标点的属性

属性	说明	类型	是否必填	备注
points	经纬度数组	Array	是	[{latitude: 0, longitude: 0}]
color	线的颜色	string	否	以8位十六进制数表示，后两位表示alpha值，如#000000AA
width	线的宽度	number	否	
dottedLine	是否虚线	boolean	否	默认为 false
arrowLine	带箭头的线	boolean	否	默认为 false，开发者工具暂不支持该属性
arrowIconPath	更换箭头图标	string	否	在 arrowLine 为 true 时生效
borderColor	线的边框颜色	string	否	
borderWidth	线的厚度	number	否	

circles 用来在地图上显示圆，如表 8.9 所示。

表 8.9 circles 显示圆的属性

属性	说明	类型	是否必填	备注
latitude	纬度	number	是	浮点数，范围为-90~90
longitude	经度	number	是	浮点数，范围为-180~180
color	描边的颜色	string	否	以8位十六进制数表示，后两位表示alpha值，如#000000AA
fillColor	填充颜色	string	否	以8位十六进制数表示，后两位表示alpha值，如#000000AA
radius	半径	number	是	
strokeWidth	描边的宽度	number	否	

controls 用来在地图上显示控件，控件不随着地图移动，如表 8.10 所示。

表 8.10 controls 显示控件的属性

属性	说明	类型	是否必填	备注
id	控件 id	number	否	在控件单击事件回调后返回此 id
position	控件在地图上的位置	Object	是	控件相对地图位置
iconPath	显示的图标	string	是	项目目录下的图片路径，支持相对路径的写法，以 "/" 开头则表示相对小程序根目录
clickable	是否可单击	boolean	否	默认不可单击

position 控件的位置是相对地图的位置，如表 8.11 所示。

表 8.11 position 控件位置的属性

属性	说明	类型	是否必填	备注
left	距离地图的左边界多远	number	否	默认为 0
top	距离地图的上边界多远	number	否	默认为 0
width	控件宽度	number	否	默认为图片宽度
height	控件高度	number	否	默认为图片高度

注意：地图组件的经纬度必填，如果不填经纬度则默认值是北京的经纬度。

示例代码如下。

```
<!-- map.wxml -->
<map id="map" longitude="113.324520" latitude="23.099994" scale="14" controls="{{controls}}" bindcontroltap="controltap" markers="{{markers}}" bindmarkertap="markertap" polyline="{{polyline}}" bindregionchange="regionchange" show-location style="width: 100%; height: 300px;"></map>
```

```
// map.js
Page({
  data: {
    markers: [{
      iconPath: "/resources/others.png",
      id: 0,
      latitude: 23.099994,
      longitude: 113.324520,
      width: 50,
      height: 50
    }],
    polyline: [{
      points: [{
        longitude: 113.3245211,
        latitude: 23.10229
      }, {
        longitude: 113.324520,
        latitude: 23.21229
      }],
      color:"#FF0000DD",
      width: 2,
      dottedLine: true
    }],
    controls: [{
      id: 1,
      iconPath: '/resources/location.png',
      position: {
        left: 0,
        top: 300 - 50,
        width: 50,
        height: 50
      },
      clickable: true
    }]
```

```
  },
  regionchange(e) {
    console.log(e.type)
  },
  markertap(e) {
    console.log(e.markerId)
  },
  controltap(e) {
    console.log(e.controlId)
  }
})
```

界面效果如图 8.2 所示。

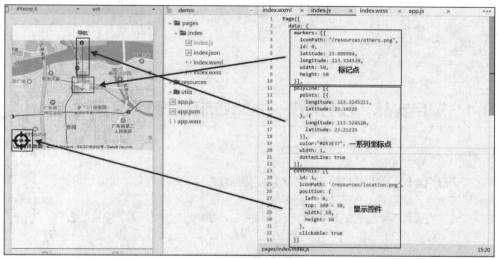

图 8.2　地图界面效果

8.3.2　地图 API 的应用

wx.createMapContext（mapId）地图组件控制 API 用来创建并返回 map 上下文 mapContext 对象，它有以下 8 种方法。

（1）MapContext.getCenterLocation（Object object）获取当前地图中心的经纬度，返回的是 gcj02 坐标系，可以用于 wx.openLocation。

（2）MapContext.moveToLocation（Object object）将地图中心移动到当前定位点，需要配合 map 组件的 show-location 使用。

（3）MapContext.getRegion（Object object）　获取当前地图的视野范围。

（4）MapContext.getRotate（Object object）　获取当前地图的旋转角。

（5）MapContext.getScale（Object object）　获取当前地图的缩放级别。

（6）MapContext.getSkew（Object object）　获取当前地图的倾斜角。

（7）MapContext.includePoints（Object object）　缩放视野展示所有经纬度。

（8）MapContext.translateMarker（Object object）　平移 marker，带动画。

示例代码如下。

```
<!-- map.wxml -->
<map id="myMap" show-location />

<button type="primary" bindtap="getCenterLocation">获取位置</button>
<button type="primary" bindtap="moveToLocation">移动位置</button>
```

```
// map.js
Page({
  onReady: function (e) {
    // 使用 wx.createMapContext 获取 map 上下文
    this.mapCtx = wx.createMapContext('myMap')
  },
  getCenterLocation: function () {
    this.mapCtx.getCenterLocation({
      success: function(res){
        console.log(res.longitude)
        console.log(res.latitude)
      }
    })
  },
  moveToLocation: function () {
    this.mapCtx.moveToLocation()
  }
})
```

8.4 项目实战：任务16——实现商品立即购买页功能

慕课视频

项目实战：实现商品立即购买页功能

1. 任务目标

通过实现商品立即购买页功能，来巩固选择收货地址设计、线上收货地址区域设计、购买商品区域设计、商品总价格区域设计的知识。

莫凡商城的立即购买页面包括选择收货地址和显示收货地址区域、购买商品区域、商品总价格区域，如图 8.3 所示。

图 8.3 立即购买页面

2. 任务实施

下面我们一起来实现莫凡商城立即购买页功能。

（1）在 app.json 文件里添加立即购买页面路径 "pages/buy/buy"。

（2）在 buy.wxml 页面文件里进行立即购买页面布局设计，示例代码如下。

```
<view class="content">
  <view class="hr"></view>
  <view class="address" bindtap="selectAddress">
    <view class="location">
      <image src="/pages/images/icon/address.png"></image>
    </view>
    <block wx:if="{{addresses != ''}}">
    <view class="desc1">
      <view>收货人:{{addresses.personName}}</view>
      <view>联系方式:{{addresses.contactnumber}}</view>
      <view>收货地址:{{addresses.city}} {{addresses.address}} {{addresses.housenumber}}</view>
    </view>
    </block>
    <block wx:else>
    <view class="addAddress">
      请选择收货地址
    </view>
    </block>
    <view class="nav"></view>
  </view>
  <view class="hr"></view>
  <view class="goods">
    <view class="title">莫凡商城</view>
    <view class="line"></view>
    <view class="good">
      <view class="pic">
        <image src="{{goodsDetail.listPic}}"></image>
      </view>
      <view class="goodInfo">
        <view class="name">{{goodsDetail.goodsName}}</view>
        <view class="price">¥{{goodsDetail.goodsPrice}}
          <text class="count">x{{num}}</text>
        </view>
      </view>
    </view>
    <view class="line"></view>
    <view class="tip">
      <view class="term">
        送货说明
      </view>
      <view class="desc2">
        <view>莫凡快递</view>
        <view>今日 22:00 前付款</view>
        <view>预计明天送达</view>
      </view>
    </view>
    <view class="line"></view>
  </view>
  <view class="hr"></view>
  <view class="bottom">
    <view class="intocart"><text>总额(含运费):<text class="total">¥{{totalPrice}}</text></text></view>
    <view class="buy" bindtap="buy">提交订单</view>
  </view>
</view>
```

（3）在 buy.wxss 样式文件里对立即购买页面进行样式渲染，示例代码如下。

```
.content{
  width: 100%;
  font-family: "Microsoft YaHei";
  background-color: #F9F9F8;
  height: 700px;
}
.hr{
  height: 10px;
}
.address{
  display: flex;
  flex-direction: row;
```

```css
    height: 100px;
    background-color: #ffffff;
    line-height: 100px;
}
.location{
    line-height:100px;
    margin-left:10px;
}
.location image{
    width: 20px;
    height: 20px;
}
.desc{
    line-height: 100px;
    padding-left: 5px;
    font-size: 15px;
    color: #ff0000;
}
.nav{
    position: absolute;
    right: 10px;
    line-height: 80px;
}
.desc1{
    padding:5px;
    font-size: 13px;
    line-height: 20px;
    align-items: center;
    font-weight: bold;
    margin-top:5px;
}
.addAddress{
    font-size: 13px;
    text-align: center;
    height: 100px;
    line-height: 100px;
    margin-left:10px;
    color: #666666;
}
.goods{
    background-color: #ffffff;
    height: 215px;
}
.title{
    padding:10px;
    font-size: 16px;
    font-weight: bold;
}
.line{
    height: 1px;
    width: 100%;
    background-color: #cccccc;
    opacity: 0.2;
}
.good{
    height: 100px;
    display: flex;
    flex-direction: row;
}
.pic{
    width: 25%;
    text-align: center;
    line-height: 100px;
}
.good image{
    width: 50px;
    height: 70px;
```

```css
    vertical-align: middle;
}
.goodInfo{
    width: 75%;
    line-height: 50px;
}
.name{
    font-size: 13px;
    font-weight: bold;
}
.price{
    color: red;
}
.count{
    margin-left: 5px;
    font-size: 12px;
    color: #999999;
}
.tip{
    display: flex;
    flex-direction: row;
    height: 70px;
}
.term{
    width: 30%;
    font-size: 13px;
    line-height: 70px;
    margin-left: 10px;
}
.desc2{
    width: 60%;
    font-size: 12px;
    text-align: right;
    margin-top:10px;
}
.bottom{
    background-color:#ffffff;
    height: 50px;
    position: fixed;
    bottom: 0px;
    width: 100%;
    display: flex;
    flex-direction: row;
}
.cart{
    width: 20%;
    height: 100%;
    text-align: center;
    line-height: 80px;
}
.intocart{
    width: 70%;
    background-color: #ffffff;
    font-size:12px;
    text-align: right;
    line-height: 50px;
    padding-right:10px;
}
.buy{
    width: 30%;
    background-color: #009966;
    color: #ffffff;
    font-size:16px;
    text-align: center;
    line-height: 50px;
}
.total{
```

```css
    color: red;
    font-size: 16px;
    font-weight: bold;
}
```

（4）在 buy.js 业务逻辑处理文件里，加载用户的收货地址、商品列表、触发选择地址事件，示例代码如下。

```javascript
var app = getApp();
var host = app.globalData.host;
Page({
    data: {
        flag: 0,
        addresses: '',
        goodsId:'',
        goodsDetail:null,
        num:1,
        addressId:'',
        totalPrice:0
    },
    onLoad: function (e) {
        console.log(e);
        var that = this;
        this.setData({goodsId: e.goodsId});
        this.setData({ num: e.num });
        this.setData({ addressId: e.addressId });
        this.loadAddress(e.addressId);
        this.loadGoods(e.goodsId);
    },
    loadAddress: function (id) {//获取用户收货地址
        var page = this;
        if(id != null && id !=""){
            wx.request({
                url: host + '/api/address/getAddressById',
                method: 'GET',
                data: {
                    "id": id
                },
                header: {
                    'Content-Type': 'application/json'
                },
                success: function (res) {
                    console.log(res);
                    var code = res.data.code;
                    var addresses = res.data.data;
                    if (code = '0000') {
                        page.setData({ addresses: addresses });
                    }
                }
            })
        }
    },
    loadGoods: function (goodsId) {//获取购买商品列表
        var page = this;
        if (goodsId != null && goodsId != "") {
            wx.request({
                url: host + '/api/goods/getGoodsDetail?goodsId=' + goodsId,
                method: 'GET',
                data: {
                    "goodsId": goodsId
                },
                header: {
                    'Content-Type': 'application/json'
                },
                success: function (res) {
                    console.log(res);
                    var code = res.data.code;
                    var goodsDetail = res.data.data;
```

```
          if (code = '0000') {
            page.setData({ goodsDetail: goodsDetail });
            var num = page.data.num;
            //计算总价格
            var totalPrice = goodsDetail.goodsPrice * num;
            page.setData({ totalPrice: totalPrice.toFixed(2) });
          }
        }
      })
    }
  },
  selectAddress: function () {//选择收货地址
    wx.navigateTo({
      url: '../address/address?goodsId=' + this.data.goodsId+"&num="+this.data.num
    })
  },
  buy: function () {//立即购买
    var userId = wx.getStorageSync("userId");
    var addressId = this.data.addressId;
    var goodsId = this.data.goodsId;
    var num = this.data.num;
    console.log(addressId + '---' + userId + '---' + goodsId+'---'+num)
    if (addressId != '' && addressId != null){
      //保存订单信息
      wx.request({
        url: host + '/api/order/saveOrder',
        method: 'GET',
        data: {
          "goodsId": goodsId,
          "userId": userId,
          "addressId": addressId,
          "num": num
        },
        header: {
          'Content-Type': 'application/json'
        },
        success: function (res) {
          console.log(res);
          var code = res.data.code;
          var orderId = res.data.data;//订单号
          if (code = '0000') {
            //支付

            //支付成功后跳转
            wx.redirectTo({
              url: '../paySuccess/paySuccess?orderId=' + orderId
            })
          }
        }
      })

    }else{
      wx.showModal({
        title: '提示',
        content: '请选择收货地址',
        showCancel:false,
        success(res) {
          if (res.confirm) {
            console.log('用户单击确定')
          } else if (res.cancel) {
            console.log('用户单击取消')
          }
        }
      })
    }
  }

})
```

8.5 项目实战：任务17——实现收货地址列表功能

慕课视频

项目实战：实现收货地址列表功能

1. 任务目标

通过实现收货地址列表功能，来巩固用户地址列表渲染设计、新增地址固定底部设计、地址编辑按钮跳转设计的知识。

莫凡商城的收货地址管理页面包括用户的地址列表、新增地址按钮、地址编辑按钮，如图 8.4 所示。

2. 任务实施

下面我们一起来实现莫凡商城收货地址列表功能。

（1）在 app.json 文件里添加地址列表页面路径"pages/address/address"。

（2）在 address.json 文件里配置地址列表导航标题"收货地址管理"，具体代码如下。

```
{
    "navigationBarTitleText": "收货地址管理"
}
```

图 8.4 地址列表页面

（3）在 address.wxml 页面文件里进行地址列表布局设计，示例代码如下。

```
<view class="content">
  <view class="hr"></view>
  <block wx:for="{{addresses}}">
    <view class="item">
      <view class="info {{flag==index?'select':'normal'}}" id="{{index}}" data-id="{{item.id}}" bindtap="switchNav">
        <view class="name">
          <text>{{item.personName}}</text>
          <text>{{item.contactnumber}}</text>
        </view>
        <view class="address">
          <text>{{item.city}}</text>
          <text>{{item.address}}</text>
          <text>{{item.housenumber}}</text>
        </view>
      </view>
      <view class="opr" bindtap='editAddress' id="{{item.id}}">
        <image src="/pages/images/icon/xg.png" style="width:33px;height:33px;"></image>
```

```
        </view>
      </view>
      <view class="line"></view>
  </block>
  <view class="bg">
      <view class="newAddress" bindtap="newAddress">+新增地址</view>
  </view>
</view>
```

（4）在 address.wxss 样式文件里对地址列表进行样式渲染，示例代码如下。

```
.content{
    font-family: "Microsoft YaHei";
    height: 700px;
    background-color: #F9F9F8;
}
.hr{
    height: 20px;
}
.item{
    background-color: #ffffff;
    display: flex;
    flex-direction: row;
    height: 75px;
    padding:10px;
}
.info{
    width:80%;
    line-height: 35px;
}
.name{
    margin-left: 20px;
    font-size: 15px;
    color:#999999;
}
.name text{
    margin-right: 10px;
}
.address{
    margin-left: 20px;
    font-size: 13px;
    color:#999999;
    line-height: 20px;
}
.address text{
    margin-right: 10px;
}
.opr{
    border-left: 1px solid #f2f2f2;
    line-height: 85px;
    width: 20%;
    text-align: center;
}
.line{
    height: 1px;
    width: 100%;
    background-color: #cccccc;
    opacity: 0.2;
}
.select{
    border-left:5px solid #009966;
}
.bg{
    background-color: #ffffff;
    height: 55px;
    border: 1px solid #f2f2f2;
    position: fixed;
    bottom: 0px;
    width: 100%;
```

```css
}
.newAddress{
    border: 1px solid #f2f2f2;
    width: 220px;
    height: 35px;
    background-color: #009966;
    line-height: 35px;
    text-align: center;
    border-radius: 5px;
    margin: 0 auto;
    margin-top:10px;
    font-size: 16px;
    color: #ffffff;
}
```

（5）在address.js业务逻辑处理文件里调用后端服务器接口获取地址列表，示例代码如下。

```js
var app = getApp();
var host = app.globalData.host;
Page({
  data:{
    flag:0,
    addresses:[],
    goodsId:'',
    num:1
  },
  onLoad:function(e){
    this.setData({ num: e.num });
    this.setData({ goodsId: e.goodsId});
    this.loadAddress();
  },
  switchNav:function(e){//选择地址
      var index = e.currentTarget.id;
      this.setData({ flag: index});
      var addressId = e.currentTarget.dataset.id
      wx.navigateTo({
        url: '../buy/buy?addressId=' + addressId + '&goodsId=' + this.data.goodsId+'&num='+this.data.num
      })
  },
  newAddress:function(e){//新增地址跳转
      wx.navigateTo({
        url: '../newAddress/newAddress?goodsId=' + this.data.goodsId + '&num=' + this.data.num
      })
  },
  editAddress: function (e) {//编辑地址跳转
      wx.navigateTo({
        url: '../newAddress/newAddress?addressId=' + e.currentTarget.id + '&goodsId=' + this.data.goodsId + '&num=' + this.data.num
      })
  },
  loadAddress:function(){//加载地址列表
    var page = this;
    var userId = wx.getStorageSync("userId");
    if (userId != "") {
        wx.request({
          url: host + '/api/address/selectAddressByUserId',
          method: 'GET',
          data: {
            "userId": userId
          },
          header: {
            'Content-Type': 'application/json'
          },
          success: function (res) {
            var code = res.data.code;
            var addresses = res.data.data;
            if (code = '0000') {
              page.setData({ addresses: addresses });
```

```
          }
        }
      })
    } else {
      wx.redirectTo({
        url: '../login/login'
      })
    }
  }
})
```

8.6 项目实战：任务 18——实现新增和编辑地址功能

1. 任务目标

通过实现新增编辑地址功能，来巩固表单相关组件的知识。

在莫凡商城收货地址管理页面中可以单击新增或者编辑地址按钮，跳转到新增编辑地址页面，页面中包括联系人、性别、手机号码、所在城市、收货地址、门牌号信息，如图 8.5 和图 8.6 所示。

图 8.5 新增地址页面

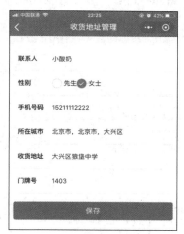

图 8.6 编辑地址页面

2. 任务实施

下面我们一起来实现莫凡商城新增编辑地址功能。

（1）在 app.json 文件里添加新增编辑地址页面路径"pages/newAddress/newAddress"。

（2）在 newAddress.json 文件里配置新增编辑地址导航标题"收货地址管理"，具体代码如下。

```
{
    "navigationBarTitleText": "收货地址管理"
}
```

（3）在 newAddress.wxml 页面文件里进行新增编辑地址布局设计，示例代码如下。

```
<view class="content">
    <view class="hr"></view>
    <view class="bg">
      <form bindsubmit="formSubmit" bindreset="formReset">
        <view class="item">
          <view class="name">联系人</view>
          <view class="value">
            <input type="text" placeholder="收货人姓名" placeholder-class="holder" name="userName" value="{{userName}}"/>
```

```
            </view>
        </view>
        <view class="line"></view>
        <view class="item">
            <view class="name">性别</view>
            <view class="value">
                <radio-group class="radin-group" bindchange="radioChange" name="sex">
                    <radio value="0" checked="{{sex==0}}">先生</radio>
                    <radio value="1" checked="{{sex==1}}">女士</radio>
                </radio-group>
            </view>
        </view>
        <view class="line"></view>
        <view class="item">
            <view class="name">手机号码</view>
            <view class="value">
                <input type="text" placeholder="联系您的电话" placeholder-class="holder" name="phone" value="{{phone}}"/>
            </view>
        </view>
        <view class="line"></view>
        <view class="item">
            <view class="name">所在城市</view>
            <view class="value">
                <picker mode="region" bindchange="bindRegionChange" value="{{region}}" custom-item="{{customItem}}" name="city">
                    <view class="picker">
                        {{region[0]}}, {{region[1]}}, {{region[2]}}
                    </view>
                </picker>
            </view>
        </view>
        <view class="line"></view>
        <view class="item">
            <view class="name">收货地址</view>
            <view class="value">
                <input type="text" placeholder="选择收货地址" placeholder-class="holder" name="address" bindtap="chooseLocation" value="{{address}}" />
            </view>
        </view>
        <view class="line"></view>
        <view class="item">
            <view class="name">门牌号</view>
            <view class="value">
                <input type="text" placeholder="请输入楼号门牌号详细信息" placeholder-class="holder" name="num" value="{{num}}"/>
            </view>
        </view>
        <view class="line"></view>
        <button class="btn" form-type="submit">保存</button>
        <view class="tip">{{tip}}</view>
    </form>
  </view>
</view>
```

（4）在 newAddress.wxss 样式文件里进行新增编辑地址样式渲染，示例代码如下。

```
.content{
    background-color: #F9F9F8;
    height: 700px;
    font-family: "Microsoft YaHei";
}
.hr{
    height: 20px;
}
.bg{
    background-color: #ffffff;
    padding:10px;
}
```

```css
.item{
    display: flex;
    flex-direction: row;
    height: 60px;
    line-height: 60px;
    align-items: center;
}
.name{
    width:20%;
    margin-left: 10px;
    font-size: 16px;
    font-weight: bold;
}
.value{
    width: 80%;
    line-height: 60px;
    margin-left: 10px;
    font-size: 16px;
}
.holder{
    color:#AEAEAE;
    font-size: 16px;
}
.line{
    border: 1px solid #cccccc;
    opacity: 0.2;
}
.btn{
    margin-top: 20px;
    background-color: #009966;
    color: #ffffff;
}
.tip{
  margin-top:10px;
  font-size: 12px;
  color: #D53E37;
  text-align: center;
}
```

（5）在 newAddress.js 业务逻辑文件里调用后端服务器接口保存地址，示例代码如下。

```javascript
var app = getApp();
var host = app.globalData.host;
Page({
  data: {
    index: 0,
    tip: '',
    address: '', //显示的地址
    region: ['北京市', '北京市', '大兴区'],
    customItem: '全部',
    addressId:'',
    sex:'',
    phone: '',
    num: '',
    userName: '',
    goodsId:'',
    goodsNum:''
  },
  onLoad: function(e) {
    this.setData({ goodsId:e.goodsId });
    this.setData({ goodsNum: e.num });
    var addressId = e.addressId;
    if (addressId != null && addressId !=''){
      this.setData({ addressId: addressId });
      this.loadAddress(addressId);
    }
  },
  loadAddress:function(addressId){
    var page = this;
```

```
        wx.request({
            url: host + '/api/address/getAddressById',
            method: 'GET',
            data: {
                "id": addressId
            },
            header: {
                'Content-Type': 'application/json'
            },
            success: function (res) {
                var code = res.data.code;
                var address = res.data.data;
                if (code = '0000') {
                    page.setData({ userName: address.personName });
                    page.setData({ sex: address.gender });
                    page.setData({ phone: address.contactnumber });
                    page.setData({ num: address.housenumber });
                    page.setData({ num: address.housenumber });
                    page.setData({ address: address.address });

                    var cities = address.city;
                    var region = cities.split(', ');
                    page.setData({ region: region });
                }
            }
        })
    },
    bindPickerChange: function(e) {
        this.setData({
            index: e.detail.value
        });
    },
    formSubmit: function(e) {
        var citys = e.detail.value.city; //所在城市
        var that = this;
        var personName = e.detail.value.userName; //联系人
        var gender = e.detail.value.sex; //性别
        var contactnumber = e.detail.value.phone; //手机号
        var address = e.detail.value.address; //收货地址
        var housenumber = e.detail.value.num; //门牌号
        var citys = e.detail.value.city; //所在城市

        var city = citys[0];
        if (citys[1] != '全部') {
            city += ',' + citys[1];
        }
        if (citys[2] != '全部') {
            city += ',' + citys[2];
        }
        var ret = that.check(personName, gender, contactnumber, address, housenumber, city);
        var userId = wx.getStorageSync("userId");
        var addressId = this.data.addressId;
        console.log('addressId=' + addressId);
        if (userId != "") {
            if (ret) {
                if (addressId != null && addressId != '') {
                    wx.request({
                        url: host + '/api/address/updateAddress',
                        method: 'GET',
                        data: {
                            "userId": userId,
                            "personName": personName,
                            "gender": gender,
                            "contactnumber": contactnumber,
                            "address": address,
                            "housenumber": housenumber,
                            "city": city,
```

```
          "addressId": addressId
        },
        header: {
          'Content-Type': 'application/json'
        },
        success: function (res) {
          var code = res.data.code;
          if (code = '0000') {
            wx.redirectTo({
              url: '../address/address?goodsId=' + that.data.goodsId + '&num=' + that.data.goodsNum
            })
          }
        }
      })
    }else{
      wx.request({
        url: host + '/api/address/saveAddress',
        method: 'GET',
        data: {
          "userId": userId,
          "personName": personName,
          "gender": gender,
          "contactnumber": contactnumber,
          "address": address,
          "housenumber": housenumber,
          "city": city
        },
        header: {
          'Content-Type': 'application/json'
        },
        success: function (res) {
          var code = res.data.code;
          if (code = '0000') {
            wx.redirectTo({
              url: '../address/address?goodsId=' + that.data.goodsId + '&num=' + that.data.goodsNum
            })
          }
        }
      })
    }
  } else {
    wx.redirectTo({
      url: '../login/login'
    })
  }
},
check: function(personName, gender, contactnumber, address, housenumber, city) {
  var that = this;
  if (personName == "") {
    that.setData({
      tip: '联系人不能为空！'
    });
    return false
  }
  if (gender == '') {
    that.setData({
      tip: '性别不能为空！'
    });
    return false
  }

  if (contactnumber == '') {
    that.setData({
      tip: '手机号不能为空！'
    });
    return false
```

```
        }
        var myreg = /^[1][3, 4, 5, 7, 8][0-9]{9}$/;
        if (!myreg.test(contactnumber)) {
          that.setData({
             tip: '手机号不合法！'
          });
          return false;
        }

        if (address == '') {
          that.setData({
             tip: '收货地址不能为空！'
          });
          return false
        }

        if (housenumber == '') {
          that.setData({
             tip: '门牌号不能为空！'
          });
          return false
        }

        if (city == '') {
          that.setData({
             tip: '所在城市不能为空！'
          });
          return false
        }

        that.setData({
           tip: ''
        });
        return true
    },
    chooseLocation: function () {
        var page = this;
        wx.chooseLocation({
           type: 'gcj02',
           success: function (res) {
              var address = res.name;
              var lat = res.latitude
              var lon = res.longitude
              page.setData({
                 address: address
              })
           }
        })
    },
    bindRegionChange: function (e) {
        console.log('picker 发送选择改变，携带值为', e.detail.value)
        this.setData({
           region: e.detail.value
        })
    }
})
```

8.7 小结

本章主要讲解获取收货地址功能的设计，这是小程序经常会用到的功能。通过本章的学习，我们综合应用获取位置相关 API、收货地址 API、地图组件和地图 API 等知识完成项目设计，在其他项目里，我们可以用同样的方式来实现收货地址的获取。

第9章 莫凡商城支付功能及订单详情页设计

微信小程序只能使用微信支付这一种支付方式。支付是小程序常用的功能,微信小程序提供支付相关 API。本章要学习设计支付功能,以及支付功能实现的整个流程。画布组件及画布 API 可用来自定义绘制一些页面,如设计分享页面,可以通过页面生成图片然后进行分享,这时就可以使用画布组件及画布 API 来实现页面生成图片的功能。

9.1 支付 API

慕课视频

支付 API

微信小程序支付功能的实现步骤如下。
(1)微信小程序调用 wx.login 方法,获取用户登录凭证 code。
(2)微信小程序将用户登录凭证 code 传输给自己的开发后台者服务器。
(3)开发者后台服务器根据用户登录凭证 code 向微信服务器请求获取唯一标识(openid)。
(4)商户后端服务器获取到唯一标识(openid)后,调用统一下单支付接口,来获取预支付交易会话标识(prepay_id)。
(5)商户后端服务器调用签名,并返回支付需要使用的参数。
(6)微信小程序调用 wx.requestPayment 方法发起微信支付。
(7)商户后端服务器接收微信服务器的通知并处理微信服务器返回的结果。

微信小程序提供了微信支付接口,可以使用 wx.requestPayment(OBJECT)来进行微信支付,具体参数说明如表 9.1 所示。

表 9.1 wx.requestPayment 的参数说明

属性	类型	是否必填	说明
timeStamp	string	是	时间戳从 1970 年 1 月 1 日 00:00:00 至今的秒数,即当前的时间
nonceStr	string	是	随机字符串,长度为 32 个字符以下
Package	string	是	统一下单接口返回的 prepay_id 参数值,提交格式如 prepay_id=*
signType	string	是	签名算法,支持 MD5、HMAC-SHA256
paySign	string	是	签名,具体签名方案参见微信公众号支付帮助文档
Success	Function	否	接口调用成功的回调函数
Fail	Function	否	接口调用失败的回调函数
complete	Function	否	接口调用结束的回调函数(调用成功、失败都会执行)

示例代码如下。

```
wx.requestPayment({
```

```
    'timeStamp': '',
    'nonceStr': '',
    'package': '',
    'signType': 'MD5',
    'paySign': '',
    'success':function(res){

    },
    'fail':function(res){
    }
})
```

9.2 项目实战：任务 19——实现支付功能

1. 任务目标

通过实现莫凡商城的支付功能，来巩固微信支付功能的相关知识。

莫凡商城在提交订单页面和订单详情页面都可以发起商品支付，计算出需要支付的总金额，发起支付。订单详情页支付在 9.5 节中实现，本节实现提交订单页发起商品支付功能，如图 9.1 所示。

图 9.1 提交订单

2. 任务实施

下面我们一起来实现莫凡商城提交订单页发起商品支付功能。在 buy.js 业务逻辑处理文件里发起微信支付，示例代码如下。

```
var app = getApp();
var host = app.globalData.host;
Page({
  data: {
    flag: 0,
    addresses: '',
    goodsId:'',
    goodsDetail:null,
```

```
    num:1,
    addressId:'',
    totalPrice:0
  },
  onLoad: function (e) {
    console.log(e);
    var that = this;
    this.setData({goodsId: e.goodsId});
    this.setData({ num: e.num });
    this.setData({ addressId: e.addressId });
    this.loadAddress(e.addressId);
    this.loadGoods(e.goodsId);
  },
  loadAddress: function (id) {//获取用户收货地址
    var page = this;
    if(id != null && id !=""){
      wx.request({
        url: host + '/api/address/getAddressById',
        method: 'GET',
        data: {
          "id": id
        },
        header: {
          'Content-Type': 'application/json'
        },
        success: function (res) {
          console.log(res);
          var code = res.data.code;
          var addresses = res.data.data;
          if (code = '0000') {
            page.setData({ addresses: addresses });
          }
        }
      })
    }
  },
  loadGoods: function (goodsId) {//获取购买商品列表
    var page = this;
    if (goodsId != null && goodsId != "") {
      wx.request({
        url: host + '/api/goods/getGoodsDetail?goodsId=' + goodsId,
        method: 'GET',
        data: {
          "goodsId": goodsId
        },
        header: {
          'Content-Type': 'application/json'
        },
        success: function (res) {
          console.log(res);
          var code = res.data.code;
          var goodsDetail = res.data.data;
          if (code = '0000') {
            page.setData({ goodsDetail: goodsDetail });
            var num = page.data.num;
            //计算总价格
            var totalPrice = goodsDetail.goodsPrice * num;
            page.setData({ totalPrice: totalPrice.toFixed(2) });
          }
        }
```

```
        })
      }
    },
    selectAddress: function () {//选择收货地址
      wx.navigateTo({
        url: '../address/address?goodsId=' + this.data.goodsId+"&num="+this.data.num
      })
    },
    buy: function () {//立即购买
      var userId = wx.getStorageSync("userId");
      var addressId = this.data.addressId;
      var goodsId = this.data.goodsId;
      var num = this.data.num;
      console.log(addressId + '---' + userId + '---' + goodsId+'---'+num)
      if (addressId != '' && addressId != null){
        //保存订单信息
        wx.request({
          url: host + '/api/order/saveOrder',
          method: 'GET',
          data: {
            "goodsId": goodsId,
            "userId": userId,
            "addressId": addressId,
            "num": num
          },
          header: {
            'Content-Type': 'application/json'
          },
          success: function (res) {
            console.log(res);
            var code = res.data.code;
            var orderId = res.data.data;//订单号
            if (code = '0000') {
              //发起支付
              var that = this;
              that.getCode();//动态获取 code
              var param = {
                "fee": that.data.totalPrice,
                "userId": wx.getStorageSync("userId"),
                "orderId": orderId,
                "appId": '0',
                "jsCode": wx.getStorageSync('jscode')
              }

              wx.request({
                url: host + '/api/pay/recharge',
                method: 'GET',
                data: { 'data': JSON.stringify(param) },
                header: {
                  'Content-Type': 'application/json'
                },
                success: function (res) {
                  var code = res.data.code;
                  var ret = res.data.data;

                  if (code == '0000') {
                    wx.requestPayment({
                      timeStamp: ret.timestamp,
                      nonceStr: ret.noncestr,
                      package: ret.package,
```

```
                signType: ret.signType,
                paySign: ret.sign,
                success(res) {
                  wx.navigateTo({//支付成功后跳转成功页
                    url: '../paySuccess/paySuccess?orderId=' + orderId
                  })
                }
              })
            } else {
              return false;
            }
          }
        })
      }
    })
  }else{
    wx.showModal({
      title: '提示',
      content: '请选择收货地址',
      showCancel:false,
      success(res) {
        if (res.confirm) {
          console.log('用户单击确定')
        } else if (res.cancel) {
          console.log('用户单击取消')
        }
      }
    })
  }
},
getCode: function () {
  wx.login({
    success: res => {
      var jscode = res.code
      wx.setStorageSync('jscode', jscode)
    }
  })
}
})
```

总的来说，发起微信支付，首先微信小程序将登录凭证 code 和订单相关信息一起提交给后端服务器，后端服务器向微信服务器发起支付，发起支付后微信服务器返回支付参数，后端服务器将支付参数返回给微信小程序，微信小程序使用这些返回参数调用 wx.requestPayment 小程序支付 API，支付成功后跳转到支付页面，来完成微信小程序支付。

9.3 画布组件及画布 API 的应用

利用 canvas 画布组件可以自定义绘制一些图像或者图形，如绘制正方形、圆形等形状，也可以将页面生成图片。

canvas 画布组件默认宽度为 300 px，高度为 225 px，在使用时需要有唯一的标识，同一页面中的 canvas-id 不可重复，如果使用一个已经出现过的 canvas-id，该 canvas 标签对应的画布将被隐藏并不再正常工作。canvas 画布组件有手指触摸动作开始、手指触摸后移动、手指触摸动作结束、手指触摸动作被打断等事件，具体属性如表 9.2 所示。

表 9.2 canvas 画布组件的属性

属性	类型	默认值	说明
type	string		指定 canvas 的类型，当前仅支持 WebGL
canvas-id	string		canvas 组件的唯一标识符
disable-scroll	boolean	false	当在 canvas 中移动时，禁止屏幕滚动及下拉刷新
bindtouchstart	eventHandle		手指触摸动作开始
bindtouchmove	eventHandle		手指触摸后移动
bindtouchend	eventHandle		手指触摸动作结束
bindtouchcancel	eventHandle		手指触摸动作被打断，如来电提醒、弹窗
bindlongtap	eventHandle		手指长按 500 ms 之后触发，触发了长按事件后进行移动不会触发屏幕的滚动
binderror	eventHandle		当发生错误时触发 error 事件,detail = {errMsg：'something wrong'}

canvas 画布组件需要和画布 API 一起使用，微信小程序提供画布相关 API 供开发者使用。可以使用 wx.createCanvasContext（string canvasId，Object this）创建 canvas 的绘图上下文 CanvasContext 对象，返回 CanvasContext 对象，CanvasContext 对象提供以下方法。

（1）CanvasContext.draw（boolean reserve, function callback）将之前在绘图上下文中的描述（路径、变形、样式）画到 canvas 中。

（2）CanvasGradient CanvasContext.createLinearGradient（number x0，number y0，number x1，number y1）创建一个线性的渐变颜色。返回的 CanvasGradient 对象需要使用 CanvasGradient.addColorStop()来指定渐变点，至少要两个。

（3）CanvasGradient CanvasContext.createCircularGradient（number x, number y, number r）创建一个圆形的渐变颜色。起点在圆心，终点在圆环。返回的 CanvasGradient 对象需要使用 CanvasGradient.addColorStop()来指定渐变点，至少要两个。

（4）CanvasContext.createPattern（string image, string repetition）对指定的图像创建模式的方法，可在指定的方向上重复元图像。

（5）Object CanvasContext.measureText（string text）测量文本尺寸信息，目前仅返回文本宽度，同步接口。

（6）CanvasContext.save()保存绘图上下文。

（7）CanvasContext.restore()恢复之前保存的绘图上下文。

（8）CanvasContext.beginPath()开始创建一个路径。需要调用 fill 或者 stroke 才会使用路径进行填充或描边，在最开始的时候相当于调用了一次 beginPath。同一个路径内的多次 setFillStyle、setStrokeStyle、setLineWidth 等设置，以最后一次设置为准。

（9）CanvasContext.moveTo（number x, number y）把路径移动到画布中的指定点，不创建线条。用 stroke 方法来画线条。

（10）CanvasContext.lineTo（number x, number y）增加一个新点，然后创建一条从上次指定点到目标点的线。用 stroke 方法来画线条。

（11）CanvasContext.quadraticCurveTo（number cpx, number cpy, number x, number y）创建二次贝塞尔曲线路径。曲线的起始点为路径中前一个点。

（12）CanvasContext.bezierCurveTo()创建三次贝塞尔曲线路径。曲线的起始点为路径中前一个点。

（13）CanvasContext.arc（number x, number y, number r, number sAngle, number eAngle, boolean counterclockwise）创建一条弧线。创建一个圆可以指定起始弧度为 0，终止弧度为 2×Math.PI。用 stroke 或者 fill 方法来在 canvas 中画弧线。

（14）CanvasContext.rect（number x, number y, number width, number height）创建一个矩形路径。需要用 fill 或者 stroke 方法将矩形真正地画到 canvas 中。

（15）CanvasContext.arcTo（number x1，number y1，number x2，number y2，number radius）根据控制点和半径绘制圆弧路径。

（16）CanvasContext.clip()从原始画布中剪切任意形状和尺寸。一旦剪切了某个区域，所有之后的绘图就都会被限制在被剪切的区域内（不能访问画布上的其他区域）。可以在使用 clip 方法前通过使用 save 方法对当前画布区域进行保存，并在以后的任意时间通过 restore 方法对其进行恢复。

（17）CanvasContext.fillRect（number x, number y, number width, number height）填充一个矩形。用 setFillStyle 设置矩形的填充色，如果未设置则默认为黑色。

（18）CanvasContext.strokeRect（number x, number y, number width, number height）画一个矩形（非填充）。用 setStrokeStyle 设置矩形线条的颜色，如果未设置则默认为黑色。

（19）CanvasContext.clearRect（number x, number y, number width, number height）清除画布上在该矩形区域内的内容。

（20）CanvasContext.fill()对当前路径中的内容进行填充。默认填充色为黑色。

（21）CanvasContext.stroke()画出当前路径的边框。默认颜色为黑色。

（22）CanvasContext.closePath()关闭一个路径，会连接起点和终点。如果关闭路径后没有调用 fill 或者 stroke 并开启了新的路径，那么之前的路径将不会被渲染。

（23）CanvasContext.scale（number scaleWidth, number scaleHeight）在调用后，之后创建的路径其横、纵坐标会被缩放。多次调用倍数会相乘。

（24）CanvasContext.rotate（number rotate）以原点为中心顺时针旋转当前坐标轴。多次调用旋转的角度会叠加。原点可以用 translate 方法修改。

（25）CanvasContext.translate（number x, number y）对当前坐标系的原点（0，0）进行变换。默认的坐标系原点为页面左上角。

（26）CanvasContext.drawImage（string imageResource, number sx, number sy, number sWidth, number sHeight, number dx, number dy, number dWidth, number dHeight）绘制图像到画布。

（27）CanvasContext.strokeText（string text, number x, number y, number maxWidth）在给定的（x, y）位置绘制文本描边的方法。

（28）CanvasContext.transform（number scaleX, number scaleY, number skewX, number skewY, number translateX, number translateY）使用矩阵多次叠加当前变换的方法。

（29）CanvasContext.setTransform（number scaleX, number scaleY, number skewX, number skewY, number translateX, number translateY）使用矩阵重新设置（覆盖）当前变换的方法。

（30）CanvasContext.setFillStyle（string|CanvasGradient color）设置填充色。

（31）CanvasContext.setStrokeStyle（string color）设置描边颜色。

（32）CanvasContext.setShadow（number offsetX, number offsetY, number blur, string color）设定阴影样式。

（33）CanvasContext.setGlobalAlpha（number alpha）设置全局画笔透明度。

（34）CanvasContext.setLineWidth（number lineWidth）设置线条的宽度。

（35）CanvasContext.setLineJoin（string lineJoin）设置线条的交点样式。

（36）CanvasContext.setLineCap（string lineCap）设置线条的端点样式。

（37）CanvasContext.setLineDash（Array.<number> pattern, number offset）设置虚线样式。

(38)CanvasContext.setMiterLimit(number miterLimit)设置最大斜接长度。斜接长度指的是在两条线交汇处内角和外角之间的距离,当CanvasContext.setLineJoin()为 miter 时才有效。

(39)CanvasContext.fillText(string text, number x, number y, number maxWidth)在画布上绘制被填充的文本。

(40)CanvasContext.setFontSize(number fontSize)设置字体的字号。

(41)CanvasContext.setTextAlign(string align)设置文字的对齐。

(42)CanvasContext.setTextBaseline(string textBaseline)设置文字的竖直对齐。

(43)OffscreenCanvas wx.createOffscreenCanvas()创建离屏 canvas 实例。

(44)wx.canvasToTempFilePath(Object object, Object this)把当前画布指定区域的内容导出生成指定大小的图片。在draw()回调里调用该方法才能保证图片导出成功。

(45)wx.canvasPutImageData(Object object, Object this)将像素数据绘制到画布。在自定义组件下,第二个参数传入自定义组件实例this,以操作组件内<canvas>组件。

(46)wx.canvasGetImageData(Object object, Object this)获取 canvas 区域隐含的像素数据。

(47)RenderingContext Canvas.getContext(string contextType)该方法返回 Canvas 的绘图上下文。

(48)Image Canvas.createImage()创建一个图片对象(仅支持在 WebGL 中使用,暂不支持在 OffscreenCanvas 中使用)。

(49)number Canvas.requestAnimationFrame(function callback)在下次进行重绘时执行(仅支持在 WebGL 中使用)。

(50)Canvas.cancelAnimationFrame(number requestID)取消由 requestAnimationFrame 添加到计划中的动画帧请求(仅支持在 WebGL 中使用)。

示例代码如下。

```
<!-- canvas.wxml -->
<canvas style="width: 300px; height: 200px;" canvas-id="firstCanvas"></canvas>
<!-- 当使用绝对定位时,文档流后边的 canvas 的显示层级高于前边的 canvas -->
<canvas style="width: 400px; height: 500px;" canvas-id="secondCanvas"></canvas>
<!-- 因为 canvas-id 与前一个 canvas 重复,该 canvas 不会显示,并会发送一个错误事件到 AppService -->
<canvas style="width: 400px; height: 500px;" canvas-id="secondCanvas" binderror="canvasIdErrorCallback">
</canvas>
```

```
// canvas.js
Page({
  canvasIdErrorCallback: function (e) {
    console.error(e.detail.errMsg)
  },
  onReady: function (e) {

    // 使用 wx.createContext 获取绘图上下文 context
    var context = wx.createContext()

    context.setStrokeStyle("#00ff00")
    context.setLineWidth(5)
    context.rect(0, 0, 200, 200)
    context.stroke()
    context.setStrokeStyle("#ff0000")
    context.setLineWidth(2)
    context.moveTo(160, 100)
    context.arc(100, 100, 60, 0, 2 * Math.PI, true)
    context.moveTo(140, 100)
```

```
            context.arc(100, 100, 40, 0, Math.PI, false)
            context.moveTo(85, 80)
            context.arc(80, 80, 5, 0, 2 * Math.PI, true)
            context.moveTo(125, 80)
            context.arc(120, 80, 5, 0, 2 * Math.PI, true)
            context.stroke()

            // 调用 wx.drawCanvas, 通过 canvasId 指定在哪张画布上绘制, 通过 actions 指定绘制行为
            wx.drawCanvas({
                canvasId: 'firstCanvas',
                actions: context.getActions() // 获取绘图动作数组
            })
        }
    })
```

9.4 项目实战：任务 20——实现支付完成页功能

慕课视频

项目实战：实现支付完成页功能

1. 任务目标

通过实现支付完成页功能，来练习支付完成后页面跳转的用法。

莫凡商城在提交订单页面和订单详情页面支付成功后，跳转到支付完成页面，支付完成页包括"支付成功"图标和"查看详情"按钮，如图 9.2 所示。

2. 任务实施

下面我们一起来实现莫凡商城支付完成页功能。

（1）在 app.json 文件里添加支付完成页面路径"pages/paySuccess/paySuccess"。

图 9.2 支付完成

（2）在 paySuccess.wxml 页面文件里进行支付完成页布局设计，示例代码如下。

```
<view>
    <view class="result_picbox">
    <image class="image"src="../images/icon/payyes.jpg"></image> </view>
    <view class="result_con">
     <text class="h5">支付成功</text>
    </view>
    <view class="botm-btn"><view class="order_btn"><button  class="foobtnrend" bindtap="orderDetail" hover-class="button-hover">查看详情</button></view></view>
</view>
```

（3）在 paySuccess.wxss 样式文件里进行支付完成页样式渲染，示例代码如下。

```
page{
    color:#525a66;
}
.result_picbox {text-align: center;padding: 65px 20px 20px;}
@media only screen and (max-width: 320px){
.result_picbox {padding-top: 30px;padding-bottom: 10px;}
}
.image { width:65%; height:100px; margin:0 auto}
.result_con {text-align: center;line-height: 20px; padding:0 20px}
.h2 {font-size: 30px;line-height: 30px;padding: 10px 0;font-weight: 400;padding-bottom: 20px;display: block;
}

.result_con p.mintxt { width:70%;font-size: 12px;color: #848c99; margin:0 auto}
.order_btn { margin: 50px auto 0 auto;    text-align: center;}
button{
```

```
    width: 90%;
    background: #009966;
    color: #fff;
}
button[disabled][type="default"], wx-button[disabled]:not([type]) {
color:rgba(0, 0, 0, 0.3);
background-color:#009966;

}
.button-hover {
    background-color: #009966;
    opacity: 0.7;
}
```

（4）在 paySuccess.js 业务逻辑处理文件里进行支付完成页逻辑处理，示例代码如下。

```
Page({
  data:{
    orderId:0
  },
  onLoad:function(e){
    var orderId = e.orderId;
    this.setData({orderId:orderId});
  },
  orderDetail:function(){
    wx.redirectTo({
      url: '../orderDetail/orderDetail?orderId='+this.data.orderId
    })
  }
})
```

9.5 项目实战：任务 8——实现我的订单功能

1. 任务目标

通过实现我的订单功能，来巩固订单列表动态渲染、订单状态页签切换的知识。

莫凡商城我的订单列表包含"待付款"列表、"待收货"列表、"已完成"列表和空列表，如图 9.3～图 9.6 所示。

图 9.3 "待付款"列表

图 9.4 "待收货"列表

第 9 章 莫凡商城支付功能及订单详情页设计

图 9.5 "已完成"列表

图 9.6 空列表

2. 任务实施

下面我们一起来实现莫凡商城我的订单功能。

（1）在 app.json 文件里添加我的订单页面路径"pages/myOrder/myOrder"。

（2）在 myOrder.wxml 页面文件里进行订单列表布局设计，示例代码如下。

```
<view class="content">
  <view class="type">
    <view class="{{currentTab==0?'select':'default'}}" data-current="0" data-status="1" bindtap="switchNav">待付款</view>
    <view class="{{currentTab==1?'select':'default'}}" data-current="1" data-status="3" bindtap="switchNav">待收货</view>
    <view class="{{currentTab==2?'select':'default'}}" data-current="2" data-status="4" bindtap="switchNav">已完成</view>
  </view>

  <view class="items">
    <view class="hr"></view>
    <swiper current="{{currentTab}}" style="height:1000px;">
      <swiper-item>
        <block wx:for="{{orders}}">
          <view class="goods">
            <view class="title">莫凡商城</view>
            <view class="line"></view>
            <view class="good" bindtap="toPay" id="{{item.id}}">
              <view class="pic">
                <image src="{{item.listPic}}"></image>
              </view>
              <view class="goodInfo">
                <view class="name">{{item.goodsName}}</view>
                <view class="price">
                  <text class="count">共{{item.num}}件商品</text> ￥{{item.payAmount}}
                </view>
              </view>
            </view>
            <view class="line"></view>
            <view class="btn">
              <text bindtap="toPay" id="{{item.id}}">去支付</text>
              <text bindtap="deleteOrder" id="{{item.id}}" data-status="1">删除订单</text>
            </view>
```

```xml
          <view class="line10"></view>
          <view class="hr"></view>
        </view>
      </block>
      <block wx:if="{{orders.length==0}}">
        <view class="gyg">
          <view>
            <image src="/pages/images/icon/default.png"></image>
          </view>
          <view class="gygbtn" bindtap="toList">
            逛一逛
          </view>
        </view>
      </block>
    </swiper-item>
    <swiper-item>
      <block wx:for="{{orders}}">
        <view class="goods" >
          <view class="title">莫凡商城</view>
          <view class="line"></view>
          <view class="good" bindtap="toBuy" id="{{item.goodsId}}">
            <view class="pic">
              <image src="{{item.listPic}}"></image>
            </view>
            <view class="goodInfo">
              <view class="name">{{item.goodsName}}</view>
              <view class="price">
                <text class="count">共{{item.num}}件商品</text>　￥{{item.payAmount}}
              </view>
            </view>
          </view>
          <view class="line"></view>
          <view class="btn">
            <text bindtap="toBuy" id="{{item.goodsId}}">再次购买</text>
            <text bindtap="deleteOrder" id="{{item.id}}" data-status="3">删除订单</text>
          </view>
          <view class="line10"></view>
          <view class="hr"></view>
        </view>
      </block>
      <block wx:if="{{orders.length==0}}">
        <view class="gyg">
          <view>
            <image src="/pages/images/icon/default.png"></image>
          </view>
          <view class="gygbtn" bindtap="toList">
            逛一逛
          </view>
        </view>
      </block>
    </swiper-item>
    <swiper-item>
      <block wx:for="{{orders}}">
        <view class="goods">
          <view class="title">莫凡商城</view>
          <view class="line"></view>
          <view class="good" bindtap="toBuy" id="{{item.goodsId}}">
            <view class="pic">
              <image src="{{item.listPic}}"></image>
            </view>
```

```
                    <view class="goodInfo">
                      <view class="name">{{item.goodsName}}</view>
                      <view class="price">
                        <text class="count">共{{item.num}}件商品</text> ￥{{item.payAmount}}
                      </view>
                    </view>
                  </view>
                  <view class="line"></view>
                  <view class="btn">
                    <text bindtap="toBuy" id="{{item.goodsId}}">再次购买</text>
                    <text bindtap="deleteOrder" id="{{item.id}}" data-status="4">删除订单</text>
                  </view>
                  <view class="line10"></view>
                  <view class="hr"></view>
                </view>
              </block>
              <block wx:if="{{orders.length==0}}">
                <view class="gyg">
                  <view>
                    <image src="/pages/images/icon/default.png"></image>
                  </view>
                  <view class="gygbtn" bindtap="toList">
                    逛一逛
                  </view>
                </view>
              </block>
            </swiper-item>
          </swiper>
        </view>
      </view>
```

（3）在myOrder.wxss样式文件里进行订单列表样式渲染，示例代码如下。

```
.content{
    font-family: "Microsoft YaHei";
    width: 100%;
}
.type{
    display: flex;
    flex-direction: row;
    width: 100%;
    margin: 0 auto;
    position: fixed;
    z-index: 999;
    background: #f2f2f2;
}
.type view{
    margin: 0 auto;
}
.select{
    font-size:16px;
    font-weight: bold;
    width: 25%;
    text-align: center;
    height: 45px;
    line-height: 45px;
    border-bottom:5rpx solid #009966;
    color: #009966;
}
.default{
    width: 25%;
    font-size:16px;
```

```
        text-align: center;
        height: 45px;
        line-height: 45px;
    }
    .hr{
        height: 12px;
        background-color: #dddddd;
    }
    .items{
      padding-top:40px;
    }
    .title{
      margin-top:10px;
      padding:10px;
      font-size: 16px;
      font-weight: bold;
    }
    .line{
      height: 1px;
        width: 100%;
        background-color: #cccccc;
        opacity: 0.2;
    }
    .line10{
        height: 1px;
        width: 100%;
        background-color: #cccccc;
        opacity: 0.2;
        margin-bottom: 10px;
    }
    .good{
      height: 100px;
      display: flex;
      flex-direction: row;
    }
    .pic{
      width: 25%;
      text-align: center;
      line-height: 100px;
    }
    .good image{
       width: 50px;
       height: 70px;
       vertical-align: middle;
    }
    .goodInfo{
       width: 75%;
       line-height: 50px;
    }
    .name{
       font-size: 13px;
       font-weight: bold;
    }
    .price{
       color: red;
       text-align: right;
       margin-right: 20px;
    }
    .count{
       margin-right: 10px;
```

```css
      font-size: 12px;
      color: #666666;
}
.tip{
    display: flex;
    flex-direction: row;
    height: 70px;
}
.term{
    width: 30%;
    font-size: 13px;
    line-height: 70px;
    margin-left: 10px;
}
.desc2{
    width: 60%;
    font-size: 12px;
    text-align: right;
    margin-top:10px;
}
.btn{
    padding: 10px;
    text-align: right;
}
.btn text{
    border: 1px solid #009966;
    padding: 3px;
    font-size: 11px;
    margin-right: 10px;
    border-radius: 5px;
}
.gyg{
    margin-top:200px;
    text-align: center;
}
.gyg image{
     width: 60px;
     height: 60px;
}
.gygbtn{
    border: 1px solid #009966;
    color: #009966;
    text-align: center;
    width: 80px;
    height: 25px;
    line-height: 25px;
    margin: 0 auto;
    font-size: 16px;
    border-radius: 5px;
}
```

（4）在 myOrder.js 业务逻辑处理文件里获取订单列表，示例代码如下。

```javascript
var app = getApp();
var host = app.globalData.host;
Page({
  data: {
    currentTab: 0,
    orders: []
  },
  onLoad: function (e) {
    var id = e.id;
```

```javascript
      var status = e.status;
      console.log(id);
      this.setData({ currentTab: id });
      this.loadOrders(status);
    },
    switchNav: function (e) {
      var page = this;
      var status = e.currentTarget.dataset.status;
      if (this.data.currentTab == e.target.dataset.current) {
        return false;
      } else {
        page.setData({ currentTab: e.target.dataset.current });
      }
      page.loadOrders(status);
    },
    toPay: function (e) {
      wx.redirectTo({
        url: '../orderDetail/orderDetail?orderId=' + e.currentTarget.id
      })
    },
    toBuy: function (e) {
      var goodsId = e.currentTarget.id;
      wx.navigateTo({
        url: '../goodsDetail/goodsDetail?goodsId=' + goodsId,
      })
    },
    toList: function (e) {
      wx.reLaunch({
        url: '../index/index'
      })
    },
    deleteOrder:function(e){
      var page = this;
      var id = e.currentTarget.id;
      var status = e.currentTarget.dataset.status;
      wx.request({
        url: host + '/api/order/deleteOrder',
        method: 'GET',
        data: {
          id: id
        },
        header: {
          'Content-Type': 'application/json'
        },
        success: function (res) {
          var code = res.data.code;
          if(code=='0000'){
            wx.showToast({
              title: '删除成功',
              icon: 'success',
              duration: 1000
            })
            page.loadOrders(status);
          }
        }
      })
    },
    loadOrders: function (orderStatus) {
      var page = this;
      var userId = wx.getStorageSync("userId");
```

```
wx.request({
    url: host + '/api/order/getOrderList',
    method: 'GET',
    data: {
        userId: userId,
        orderStatus, orderStatus
    },
    header: {
        'Content-Type': 'application/json'
    },
    success: function (res) {
        var orders = res.data.data;
        console.log(orders);
        page.setData({
            orders: orders
        });
    }
})
```

9.6 项目实战：任务 21——实现订单详情页功能

慕课视频

项目实战：实现订单详情页功能

1. 任务目标

通过实现订单详情页功能，来巩固订单详情页布局设计的相关知识。

莫凡商城订单详情页用来显示已经下单的商品，该订单如果支付成功就不再显示"去付款"按钮，否则显示"去付款"按钮，如图 9.7 所示。

图 9.7 订单详情

2. 任务实施

下面我们一起来实现莫凡商城订单详情页功能。

（1）在 app.json 文件里添加订单详情页面路径 "pages/orderDetail/orderDetail"。
（2）在 orderDetail.wxml 页面文件里进行订单详情页布局设计，示例代码如下。

```
<view class="content">
  <view class="hr"></view>
  <view class="order">
    <view class='title'>
      <text>订单编号:{{orderDetail.id}}</text>
      <text class="orderStatus" wx:if="{{orderDetail.orderStatus == 1}}">待付款</text>
      <text class="orderStatus" wx:elif="{{orderDetail.orderStatus == 1}}">待发货</text>
      <text class="orderStatus" wx:elif="{{orderDetail.orderStatus == 2}}">待收货</text>
      <text class="orderStatus" wx:elif="{{orderDetail.orderStatus == 3}}">交易成功</text>
      <text class="orderStatus" wx:elif="{{orderDetail.orderStatus == 4}}">退款</text>
      <text class="orderStatus" wx:elif="{{orderDetail.orderStatus == 5}}">交易关闭</text>
    </view>
    <view class="line"></view>
    <view class='item'>
      <text>商品单价</text>
      <text class="orderStatus">¥ {{goodsDetail.goodsPrice}}</text>
    </view>
    <view class='item'>
      <text>商品数量</text>
      <text class="orderStatus"> x {{num}}</text>
    </view>
    <view class='item'>
      <text>运费(快递)</text>
      <text class="orderStatus">¥0.00</text>
    </view>
    <view class='item'>
      <text>订单总价</text>
      <text class="orderStatus">¥ {{totalPrice}}</text>
    </view>
    <view class='item'>
      <text>创建时间</text>
      <text class="orderStatus">{{orderDetail.createTime}}</text>
    </view>
  </view>
  <view class="hr"></view>
  <view class="address">
    <view class="location">
      <image src="/pages/images/icon/address.png"></image>
    </view>
    <view class="desc1">
      <view>{{addresses.personName}} {{addresses.contactnumber}} </view>
      <view>{{addresses.city}} {{addresses.address}} {{addresses.housenumber}} </view>
    </view>
  </view>
  <view class="hr"></view>
  <view class="goods">
    <view class="title">莫凡商城</view>
    <view class="line"></view>
    <view class="good">
      <view class="pic">
        <image src="{{goodsDetail.listPic}}"></image>
      </view>
      <view class="goodInfo">
        <view class="name">{{goodsDetail.goodsName}}</view>
        <view class="price">¥{{goodsDetail.goodsPrice}}
          <text class="count">x{{num}}</text>
        </view>
      </view>
    </view>
```

```
        </view>
        <view class="line"></view>
        <view class="tip">
            <view class="term">
                送货说明
            </view>
            <view class="desc2">
                <view>莫凡快递</view>
                <view>今日 22:00 前付款</view>
                <view>预计明天送达</view>
            </view>
        </view>
        <view class="line"></view>
    </view>
    <view class="hr"></view>
    <block wx:if="{{orderDetail.payStatus==0}}">
        <view class="bottom">
            <view class="intocart">
                <text>总额(含运费):<text class="total">¥{{totalPrice}}</text></text>
            </view>
            <view class="buy" bindtap="buy">去付款</view>
        </view>
    </block>
</view>
```

（3）在 orderDetail.wxss 样式文件里进行订单详情页样式渲染，示例代码如下。

```
.content{
    width: 100%;
    font-family: "Microsoft YaHei";
    background-color: #F9F9F8;
    height: 600px;
}
.hr{
    height: 10px;
}
.address{
    display: flex;
    flex-direction: row;
    height: 60px;
    background-color: #ffffff;
}
.location{
    line-height:60px;
    margin-left:10px;
}
.location image{
    width: 20px;
    height: 20px;
}
.desc{
    line-height: 80px;
    padding-left: 5px;
    font-size: 15px;
    color: #ff0000;
}
.nav{
    position: absolute;
    right: 10px;
    line-height: 80px;
}
.desc1{
```

```css
    padding:5px;
    font-size: 13px;
    line-height: 20px;
    align-items: center;
    font-weight: bold;
    margin-top:5px;
}
.goods{
    background-color: #ffffff;
    height: 215px;
}
.order{
    background-color: #ffffff;
    height: 190px;
}
.orderNum{
    margin: 10px;
}
.orderStatus{
    margin-right:10px;
    float: right;
    font-size: 12px;
    font-weight: normal;
    color:red;
}
.title{
    padding:10px;
    font-size: 16px;
    font-weight: bold;
}
.line{
    height: 1px;
    width: 100%;
    background-color: #cccccc;
    opacity: 0.2;
}
.good{
    height: 100px;
    display: flex;
    flex-direction: row;
}
.pic{
    width: 25%;
    text-align: center;
    line-height: 100px;
}
.good image{
    width: 50px;
    height: 70px;
    vertical-align: middle;
}
.goodInfo{
    width: 75%;
    line-height: 50px;
}
.name{
    font-size: 13px;
    font-weight: bold;
}
.price{
```

```css
    color: red;
}
.count{
    margin-left: 5px;
    font-size: 12px;
    color: #999999;
}
.tip{
    display: flex;
    flex-direction: row;
    height: 70px;
}
.term{
    width: 30%;
    font-size: 13px;
    line-height: 70px;
    margin-left: 10px;
}
.desc2{
    width: 60%;
    font-size: 12px;
    text-align: right;
    margin-top:10px;
}
.bottom{
    background-color:#ffffff;
    height: 50px;
    position: fixed;
    bottom: 0px;
    width: 100%;
    display: flex;
    flex-direction: row;
}
.cart{
    width: 20%;
    height: 100%;
    text-align: center;
    line-height: 80px;
}
.intocart{
    width: 70%;
    background-color: #ffffff;
    font-size:12px;
    text-align: right;
    line-height: 50px;
    padding-right:10px;
}
.buy{
    width: 30%;
    background-color: #009966;
    color: #ffffff;
    font-size:16px;
    text-align: center;
    line-height: 50px;
}
.total{
    color: red;
    font-size: 16px;
    font-weight: bold;
}
```

```css
.item{
    margin: 10px;
    font-size: 13px;
}
```

（4）在 orderDetail.js 业务逻辑处理文件里进行订单详情页业务逻辑处理，示例代码如下。

```javascript
var app = getApp();
var host = app.globalData.host;
Page({
  data: {
    flag: 0,
    addresses: '',
    goodsId: '',
    goodsDetail: null,
    num: 1,
    addressId: '',
    totalPrice: 0,
    orderDetail:null,
    orderId: null
  },
  onLoad: function (e) {
    var orderId = e.orderId;
    this.setData({ orderId: orderId });
    this.loadOrder(orderId);
  },
  loadOrder:function(orderId){
    var page = this;
    if (orderId != null && orderId != "") {
      wx.request({
        url: host + '/api/order/getOrderById',
        method: 'GET',
        data: {
          "orderId": orderId
        },
        header: {
          'Content-Type': 'application/json'
        },
        success: function (res) {
          console.log(res);
          var code = res.data.code;
          var orderDetail = res.data.data;
          if (code = '0000') {
            var addressId = orderDetail.addressId;
            page.setData({ orderDetail: orderDetail });
            page.loadAddress(orderDetail.addressId);
            page.loadGoods(orderDetail.goodsId);
            page.setData({ goodsId: orderDetail.goodsId });
            page.setData({ num: orderDetail.num });
            page.setData({ addressId: orderDetail.addressId });
          }
        }
      })
    }
  },
  loadAddress: function (id) {
    var page = this;
    if (id != null && id != "") {
      wx.request({
        url: host + '/api/address/getAddressById',
        method: 'GET',
        data: {
```

```javascript
          "id": id
        },
        header: {
          'Content-Type': 'application/json'
        },
        success: function (res) {
          console.log(res);
          var code = res.data.code;
          var addresses = res.data.data;
          if (code = '0000') {
            page.setData({ addresses: addresses });
          }
        }
      })
    }
  },
  loadGoods: function (goodsId) {
    var page = this;
    if (goodsId != null && goodsId != "") {
      wx.request({
        url: host + '/api/goods/getGoodsDetail?goodsId=' + goodsId,
        method: 'GET',
        data: {
          "goodsId": goodsId
        },
        header: {
          'Content-Type': 'application/json'
        },
        success: function (res) {
          console.log(res);
          var code = res.data.code;
          var goodsDetail = res.data.data;
          if (code = '0000') {
            page.setData({ goodsDetail: goodsDetail });
            var num = page.data.num;
            //计算总价格
            var totalPrice = goodsDetail.goodsPrice * num;
            page.setData({ totalPrice: totalPrice.toFixed(2) });
          }
        }
      })
    }
  },
  buy: function () {
    var that = this;
    that.getCode();//动态获取 code
    var param = {
      "fee": that.data.realPrice,
      "userId": wx.getStorageSync("userId"),
      "orderId": this.data.orderId,
      "appId": '0',
      "jsCode": wx.getStorageSync('jscode')
    }

    wx.request({
      url: host + '/api/pay/recharge',
      method: 'GET',
      data: { 'data': JSON.stringify(param) },
      header: {
        'Content-Type': 'application/json'
```

```
      },
      success: function (res) {
        var code = res.data.code;
        var ret = res.data.data;

        if (code == '0000') {
          wx.requestPayment({
            timeStamp: ret.timestamp,
            nonceStr: ret.noncestr,
            package: ret.package,
            signType: ret.signType,
            paySign: ret.sign,
            success(res) {
              wx.navigateTo({
                url: '../paySuccess/paySuccess?orderId=' + this.data.orderId
              })
            }
          })
        } else {
          return false;
        }
      }
    })
  },
  getCode: function () {
    wx.login({
      success: res => {
        var jscode = res.code
        wx.setStorageSync('jscode', jscode)
      }
    })
  }
})
```

9.7 小结

本章讲解微信小程序支付功能的使用，微信小程序提供 wx.requestPayment 支付 API，通过此支付 API 和后端服务器接口来实现支付功能，支付完成后跳转到支付完成页面。本章同时讲解使用画布组件及画布 API 来自定义绘制页面，如实现页面生成图片功能，并实现分享功能。

第10章
小程序扩展应用

微信小程序提供了丰富的组件和丰富的 API，一个完整的项目不能把所有的组件和 API 都应用到。本章介绍一些本案例没有应用到的 API 接口，包括设备应用相关 API 接口、文件操作相关 API 接口、窗口 API 接口、微信运动 API 接口，这些接口在不同的项目中可以应用到。

10.1 设备应用 API

微信小程序提供设备应用相关的 API，包括获得系统信息、获取网络状态、加速度计、罗盘、拨打电话、扫码、剪贴板、蓝牙、屏幕亮度、震动、手机联系人等。

10.1.1 获得系统信息

获得系统信息提供了两个 API：一个是异步获取系统信息的 wx.getSystemInfo（OBJECT）；另一个是同步获取系统信息的 wx.getSystemInfoSync()。success 返回参数说明如表 10.1 所示。

表 10.1　success 返回参数说明

参数	说明
brand	设备品牌
model	手机型号
pixelRatio	设备像素比
screenWidth	屏幕宽度，单位为像素
screenHeight	屏幕高度，单位为像素
windowWidth	可使用窗口宽度，单位为像素
windowHeight	可使用窗口高度，单位为像素
statusBarHeight	状态栏的高度，单位为像素
language	微信设置的语言
version	微信版本号
system	操作系统及版本
platform	客户端平台
fontSizeSetting	用户字体大小（单位为像素）。以微信客户端"我"→"设置"→"通用"→"字体大小"中的设置为准
SDKVersion	客户端基础库版本
benchmarkLevel	设备性能等级（仅 Android 小游戏）。取值为-2 或 0（该设备无法运行小游戏），-1（性能未知），>=1（设备性能值，该值越高，设备性能越好，目前最高不到 50）
albumAuthorized	允许微信使用相册的开关（仅 iOS 有效）
cameraAuthorized	允许微信使用摄像头的开关
locationAuthorized	允许微信使用定位的开关

续表

参数	说明
MicrophoneAuthorized	允许微信使用麦克风的开关
notificationAuthorized	允许微信通知的开关
notificationAlertAuthorized	允许微信通知带有提醒的开关（仅 iOS 有效）
notificationBadgeAuthorized	允许微信通知带有标记的开关（仅 iOS 有效）
notificationSoundAuthorized	允许微信通知带有声音的开关（仅 iOS 有效）
bluetoothEnabled	蓝牙的系统开关
locationEnabled	地理位置的系统开关
wifiEnabled	Wi-Fi 的系统开关
safeArea	在竖屏正方向下的安全区域

使用 wx.getSystemInfo（OBJECT）可以异步获取系统信息，示例代码如下。

```
Page({
  onLoad:function(){
    wx.getSystemInfo({
      success: function(res) {
        console.log("手机型号="+res.model)
        console.log("设备像素比="+res.pixelRatio)
        console.log("窗口宽度="+res.windowWidth)
        console.log("窗口高度="+res.windowHeight)
        console.log("微信设置的语言="+res.language)
        console.log("微信版本号="+res.version)
        console.log("操作系统版本="+res.system)
        console.log("客户端平台="+res.platform)
      }
    })
  }
})
```

使用 wx.getSystemInfoSync 可以同步获取系统信息，它是没有参数的，示例代码如下。

```
Page({
  onLoad: function () {
    try {
      var res = wx.getSystemInfoSync()
      console.log("手机型号=" + res.model)
      console.log("设备像素比=" + res.pixelRatio)
      console.log("窗口宽度=" + res.windowWidth)
      console.log("窗口高度=" + res.windowHeight)
      console.log("微信设置的语言=" + res.language)
      console.log("微信版本号=" + res.version)
      console.log("操作系统版本=" + res.system)
      console.log("客户端平台=" + res.platform)
    } catch (e) {
      // Do something when catch error
    }
  }
})
```

10.1.2 获取网络状态

微信小程序使用 wx.getNetworkType（OBJECT）来获取网络类型，网络类型分为 2g（2G 网络）、3g（3G 网络）、4g（4G 网络）、wifi（Wi-Fi 网络）、unknown（Android 下不常见的网络类型）、none（无网络）。示例代码如下。

```
Page({
  onLoad: function () {
```

```
wx.getNetworkType({
    success: function (res) {
        // 返回网络类型 2g, 3g, 4g, wifi
        var networkType = res.networkType;
        console.log("网络类型="+networkType);
    }
  })
})
```

微信小程序使用 wx.onNetworkStatusChange（function callback）监听网络状态变化事件，返回 isConnected 当前是否有网络连接、workType 网络类型。wx.offNetworkStatusChange（function callback）可以取消监听网络状态变化事件。

示例代码如下。

```
wx.onNetworkStatusChange(function (res) {
    console.log(res.isConnected)
    console.log(res.networkType)
})
```

10.1.3 加速度计

微信小程序使用 wx.onAccelerometerChange（CALLBACK）来实现加速度计，监听加速度计的数据，频率为5次/s。使用 wx.offAccelerometerChange（function callback）可以取消监听加速度计的数据事件，具体参数如表10.2所示。

表10.2 wx.onAccelerometerChange 的参数说明

参数	类型	说明
X	number	X 轴
Y	number	Y 轴
Z	number	Z 轴

示例代码如下。
```
Page({
    onLoad: function () {
        wx.onAccelerometerChange(function(res) {
            console.log("X 轴="+res.x)
            console.log("Y 轴="+res.y)
            console.log("Z 轴="+res.z)
        })
    }
})
```

微信小程序使用 wx.startAccelerometer（OBJECT）来开始监听加速度计，使用 wx.stopAccelerometer（OBJECT）来停止监听加速度计。

10.1.4 罗盘

微信小程序使用 wx.onCompassChange（CALLBACK）来监听罗盘数据，频率为5次/秒，返回值参数 direction 为面对的方向度数。可使用 wx.offCompassChange（function callback）来取消监听罗盘数据变化事件。

微信小程序开始监听罗盘数据使用 wx.startCompass（OBJECT），停止监听罗盘数据使用 wx.stopCompass（OBJECT）。

示例代码如下。
```
Page({
```

```
onLoad: function () {
    wx.startCompass() //开始监听罗盘数据
    wx.onCompassChange(function (res) {
        console.log("面对的方向度数="+res.direction)
    })
    wx.stopCompass() //停止监听罗盘数据
  }
})
```

10.1.5 拨打电话

微信小程序使用 wx.makePhoneCall（OBJECT）来拨打电话，参数 phonenumber 为需要拨打的电话号码。

示例代码如下。

```
wx.makePhoneCall({
  phonenumber: '15112345678'
})
```

10.1.6 扫码

微信小程序使用 wx.scanCode（OBJECT）来调取客户端扫码界面，扫码成功后返回对应的结果，具体参数如表 10.3 所示。

表 10.3 wx.scanCode 的参数说明

属性	类型	是否必填	说明
onlyFromCamera	boolean	否	是否只能从相机扫码，不允许从相册选择图片
scanType	Array.<string> ['barCode', 'qrCode']	否	扫码类型：barCode 为一维码、qrCode 为二维码、datamatrix 为 Data Matrix 码、pdf417 为 PDF417 条码
success	function	是	接口调用成功的回调函数，返回内容详见返回参数说明
fail	function	否	接口调用失败的回调函数
complete	function	否	接口调用结束的回调函数（调用成功、失败都会执行）

success 返回参数说明如表 10.4 所示。

表 10.4 success 返回参数说明

参数	说明
result	扫码的内容
scanType	所扫码的类型
charSet	所扫码的字符集
path	当所扫的码为当前小程序二维码时，会返回此字段，内容为二维码携带的 path
rawData	原始数据，base64 编码

示例代码如下。

```
wx.scanCode({
  success: (res) => {
    console.log(res)
  }
})
```

10.1.7 剪贴板

微信小程序提供剪贴板的功能，可以使用 wx.setClipboardData（OBJECT）设置剪贴板的内容，同时可以使用 wx.getClipboardData（OBJECT）来获取剪贴板的内容，使用 wx.setClipboardData（OBJECT）设置剪贴板的内容。

示例代码如下。

```
Page({
    onLoad: function () {
        wx.setClipboardData({
            data: '我是剪贴板内容',
            complete: function (res) {
                wx.getClipboardData({
                    success: function (res) {
                        console.log(res.data)
                    }
                })
            }
        })
    }
})
```

10.1.8 蓝牙

微信小程序针对蓝牙功能提供了很多 API，包括初始化蓝牙功能、关闭蓝牙功能、监听蓝牙功能、搜寻附近蓝牙设备等 API。

（1）wx.openBluetoothAdapter（OBJECT）用来初始化小程序的蓝牙功能，生效周期为调用 wx.openBluetoothAdapter 至调用 wx.closeBluetoothAdapter，或直至小程序被销毁为止。在小程序蓝牙适配器模块生效期间，开发者可以正常调用下面的小程序 API，并会收到蓝牙模块相关的 on 回调。

（2）wx.closeBluetoothAdapter（OBJECT）用来关闭蓝牙模块，使其进入未初始化状态。调用该方法将断开所有已建立的链接并释放系统资源。建议在使用小程序蓝牙流程后调用，与 wx.openBluetoothAdapter 成对调用。

（3）wx.onBluetoothAdapterStateChange（CALLBACK）用来监听蓝牙状态，返回值 available 为 true 代表蓝牙适配器可用、discovering 为 true 代表蓝牙适配器处于搜索状态。

（4）wx.getBluetoothAdapterState（OBJECT）用来获取蓝牙状态，返回值为 discovering 代表正在搜索设备，返回值为 available 代表蓝牙适配器可用。

（5）wx.startBluetoothDevicesDiscovery（OBJECT）用来开始搜寻附近的蓝牙外围设备。注意，该操作比较耗费系统资源，需在搜索并连接到设备后调用 wx.stopBluetoothDevicesDiscovery（OBJECT）停止搜寻附近的蓝牙外围设备。

（6）wx.getBluetoothDevices（OBJECT）用来获取在小程序蓝牙模块生效期间所有已发现的蓝牙设备，包括已经和本机处于连接状态的设备。wx.getConnectedBluetoothDevices（OBJECT）用来根据 uuid 获取处于已连接状态的设备。wx.onBluetoothDeviceFound（CALLBACK）用来监听寻找到新设备的事件。

（7）wx.createBLEConnection（OBJECT）用来连接低功耗蓝牙设备；wx.closeBLEConnection（OBJECT）用来断开与低功耗蓝牙设备的连接；wx.onBLEConnectionStateChange（CALLBACK）用来监听低功耗蓝牙连接状态的改变事件，包括开发者主动连接或断开连接、设备丢失、连接异常断开等；wx.notifyBLECharacteristicValueChange（OBJECT）用来启用低功耗蓝牙设备特征值变化时的 notify 功能，订阅特征值，设备的特征值必须支持 notify 或者 indicate 才可以成功调用，具体参照 characteristic 的 properties 属性；wx.onBLECharacteristicValueChange（CALLBACK）用来监听低功耗蓝牙设备的特征值变化，必须先启用 notify 接口才能接收到设备推送的 notification；wx.readBLECharacteristicValue（OBJECT）用来读取低功耗蓝牙设备特征值的二进制数据值；wx.writeBLECharacteristicValue（OBJECT）用来向低功耗蓝牙设备特征值中写入二进制数据。

（8）wx.getBLEDeviceServices（OBJECT）用来获取蓝牙设备所有 service（服务）；wx.getBLEDeviceCharacteristics（OBJECT）用来获取蓝牙设备某个服务中的所有 characteristic（特征值）。

10.1.9 屏幕亮度

wx.setScreenBrightness（OBJECT）用来设置屏幕亮度，它有一个参数值 value，范围是 0~1，0 代表最暗，1 代表最亮；wx.getScreenBrightness（OBJECT）用来获取屏幕的亮度；wx.setKeepScreenOn（OBJECT）用来设置是否保持常亮状态，仅在当前小程序生效，离开小程序后设置失效，它有一个参数 keepScreenOn，表示是否保持屏幕长亮；wx.onUserCaptureScreen（function callback）用来监听用户主动截屏事件，用户使用系统截屏按键截屏时触发该监听事件，并且该监听事件只能注册一个；wx.offUserCaptureScreen（function callback）用来取消监听用户主动截屏事件。

10.1.10 震动

微信小程序使用 wx.vibrateLong（OBJECT）使手机发生较长时间的振动（400 ms）；使用 wx.vibrateShort（OBJECT）使手机发生较短时间的振动（15 ms）。

10.1.11 手机联系人

微信小程序使用 wx.addPhoneContact（OBJECT）调用表单后，用户可以选择将该表单以"新增联系人"或"添加到已有联系人"的方式，写入手机系统通讯录，完成手机通讯录联系人和联系方式的增加。具体参数如表 10.5 所示。

表 10.5 wx.addPhoneContact（OBJECT）的参数说明

属性	类型	是否必填	说明
photoFilePath	string	否	头像本地文件路径
nickName	string	否	昵称
lastName	string	否	姓氏
middleName	string	否	中间名
firstName	string	是	名字
Remark	string	否	备注
mobilePhonenumber	string	否	手机号
weChatnumber	string	否	微信号
addressCountry	string	否	联系地址国家
addressState	string	否	联系地址省份
addressCity	string	否	联系地址城市
addressStreet	string	否	联系地址街道
addressPostalCode	string	否	联系地址邮政编码
Organization	string	否	公司

续表

属性	类型	是否必填	说明
Title	String	否	职位
workFaxnumber	string	否	工作传真
workPhonenumber	string	否	工作电话
hostnumber	string	否	公司电话
Email	string	否	电子邮件
url	string	否	网站
workAddressCountry	string	否	工作地址国家
workAddressState	string	否	工作地址省份
workAddressCity	string	否	工作地址城市
workAddressStreet	string	否	工作地址街道
workAddressPostalCode	string	否	工作地址邮政编码
homeFaxnumber	string	否	住宅传真
homePhonenumber	string	否	住宅电话
homeAddressCountry	string	否	住宅地址国家
homeAddressState	string	否	住宅地址省份
homeAddressCity	string	否	住宅地址城市
homeAddressStreet	string	否	住宅地址街道
homeAddressPostalCode	string	否	住宅地址邮政编码
Success	Function	否	接口调用成功的回调函数，返回内容详见返回参数说明
Fail	Function	否	接口调用失败的回调函数
Complete	Function	否	接口调用结束的回调函数（调用成功、失败都会执行）

示例代码如下。

```
Page({
  onLoad:function(){
    wx.addPhoneContact({
      firstName: '名字',
      nickName:'昵称',
      lastName:'姓氏',
      mobilePhonenumber:'手机号'
    })
  }
})
```

慕课视频

文件操作 API

10.2 文件操作 API

微信小程序提供了针对文件操作的 API，包括 wx.saveFile 保存文件到本地、wx.getSavedFileList 获取本地文件列表、wx.getSavedFileInfo 获取本地文件信息、wx.removeSavedFile 删除本地文件、wx.openDocument 打开文档、wx.getFileInfo 获取文件信息等。

10.2.1 wx.saveFile 保存文件到本地

wx.saveFile（object）可以用来根据文件的临时路径，将文件保存到本地，下次启动微信小程序的时候，仍然可以获取到该文件。如果是临时路径，下次启动微信小程序的时候，就无法获取到该文件了。本地文件存储的大小限制为 10 MB。wx.saveFile 可用来移动临时文件，因此调用成功后传入的 tempFilePath 将不可用，属性参数 tempFilePath 为需要保存的文件的临时路径。

示例代码如下。

```
Page({
  onLoad:function(){
    wx.getImageInfo({
      src: 'https://ss2.bdstatic.com/70cFvnSh_Q1YnxGkpoWK1HF6hhy/it/u=49292017,22064401&fm=28&gp=0.jpg',
      success: function (res) {
        var path = res.path;
        console.log("临时文件路径="+path);
        wx.saveFile({
          tempFilePath: path,
          success: function(res){
            var savedFilePath = res.savedFilePath;
            console.log("本地文件路径="+savedFilePath);
          }
        })
      }
    })
  }
})
```

将文件保存到本地后，会返回一个 savedFilePath 本地文件存储路径，根据这个路径可以访问或者使用该文件。在微信小程序下次启动的时候，这个本地文件仍然存在。

10.2.2　wx.getSavedFileList 获取本地文件列表

通过 wx.saveFile 可以将临时文件保存到本地，使之成为本地文件。用 wx.getSavedFileList 来获取本地文件列表，可以获取到 wx.saveFile 保存的文件，调用成功后返回 errMsg 接口调用结果和 fileList 文件列表，fileList 文件列表如表 10.6 所示。

表 10.6　fileList 说明

键	类型	说明
filePath	string	文件的本地路径
createTime	number	文件保存时的时间戳，从 1970/01/01 08：00：00 到当前时间的秒数
size	number	文件大小，单位为字节

示例代码如下。

```
Page({
  onLoad:function(){
    wx.getSavedFileList({
      success: function(res) {
        var fileList = res.fileList;
        console.log(fileList)
        for(var i=0;i<fileList.length;i++){
          var file = fileList[i];
          console.log("第"+(i+1)+"个文件:");
          console.log("文件创建时间="+file.createTime);
          console.log("文件大小="+file.size);
          console.log("文件本地路径="+file.filePath);
        }
      }
    })
  }
})
```

10.2.3　wx.getSavedFileInfo 获取本地文件信息

wx.getSavedFileInfo 用于获取本地文件的文件信息。此接口只能用于已保存到本地的文件，若需要获取临时文件信息，需使用 wx.getFileInfo()接口。通过 wx.getSavedFileInfo 获取的本地指定路径的文件信息包括文件的创建时间、文件大小及接口调用结果。wx.getSavedFileInfo 的属性参数 filePath 代表文件路径。

示例代码如下。

```
Page({
    onLoad:function(){
        wx.getSavedFileList({
            success: function(res) {
                var fileList = res.fileList;
                console.log(fileList)
                var file = fileList[0];
                wx.getSavedFileInfo({
                    filePath: file.filePath,
                    success: function(res){
                        console.log("文件创建时间="+res.createTime);
                        console.log("文件大小="+res.size);
                        console.log("文件本地路径="+res.errMsg);
                    }
                })
            }
        })
    }
})
```

10.2.4　wx.removeSavedFile 删除本地文件

wx. removeSavedFile 用来删除本地文件，属性参数 filePath 为需要删除的文件路径。示例代码如下。

```
Page({
    onLoad:function(){
        wx.getSavedFileList({
            success: function(res) {
                var fileList = res.fileList;
                console.log(fileList)
                var file = fileList[0];
                wx.removeSavedFile({
                    filePath: file.filePath,
                    complete: function(res) {
                        console.log(res)
                    }
                })
            }
        })
    }
})
```

10.2.5　wx.openDocument 打开文档

wx.openDocument 可以用来打开 doc、docx、xls、xlsx、ppt、pptx、pdf 等多种格式的文档，wx.openDocument 的属性参数 filePath 为文件路径，可通过 downFile 获得。

示例代码如下。

```
Page({
    onLoad:function(){
        wx.downloadFile({
            url: 'http://www.crcc.cn/portals/0/word/应聘材料样本.doc',
            success: function (res) {
```

```
            var filePath = res.tempFilePath
        wx.openDocument({
          filePath: filePath,
          success: function (res) {
            console.log('打开文档成功')
          }
        })
      }
    }
  })
```

10.2.6　wx.getFileInfo 获取文件信息

wx.getFileInfo（Object object）用来获取文件信息，属性参数 filePat 为本地文件路径、digestAlgorithm 为计算文件摘要的算法（md5 算法、sha1 算法），返回值为 size 文件大小，以字节为单位，digest 按照传入的 digestAlgorithm 计算得出文件摘要。

示例代码如下。

```
wx.getFileInfo({
  success (res) {
    console.log(res.size)
    console.log(res.digest)
  }
})
```

慕课视频

窗口 API

10.3　窗口 API

微信小程序提供窗口 API，可以使用 wx.onWindowResize（function callback）监听窗口尺寸变化事件，使用 wx.offWindowResize（function callback）取消监听窗口尺寸变化事件。

wx.onWindowResize（function callback）窗口返回值为 size 对象，size 对象包含两个属性：windowWidth 变化后的窗口宽度，单位为像素；windowHeight 变化后的窗口高度，单位为像素。

示例代码如下。

```
wx.onWindowResize(function (res) {
console.log(res.size. windowWidth);
console.log(res.size. windowHeight);
  });
```

慕课视频

微信运动 API

10.4　微信运动 API

微信小程序提供微信运动 wx.getWeRunData（Object object）接口 API，调用前需要用户授权 scope.werun，根据微信运动 API 获取用户过去 30 天的微信运动步数，需要先调用 wx.login 接口，步数信息会在用户主动进入小程序时更新。

微信运动 API 的返回值如下。

（1）encryptedData：包括敏感数据在内的完整用户信息的加密数据，详见加密数据解密算法。

（2）iv：加密算法的初始向量。

（3）cloudID：敏感数据对应的云 ID，只有开通了云开发的小程序才会返回，可通过云调用直接获取开放数据。

示例代码如下。

```
wx.getWeRunData({
  success (res) {
```

```
      // 拿 encryptedData 到开发者后台解密开放数据
      const encryptedData = res.encryptedData
      // 或拿 cloudID 通过云调用直接获取开放数据
      const cloudID = res.cloudID
    }
  })
```
开放数据 JSON 结构如下。
```
{
  "stepInfoList": [
    {
      "timestamp": 1445866601,
      "step": 100
    },
    {
      "timestamp": 1445876601,
      "step": 120
    }
  ]
}
```
Timestamp 为时间戳，表示数据对应的时间，step 表示微信运动步数。

10.5 项目实战：任务 22——实现图书分类功能

1. 任务目标

通过实现图书分类功能，来巩固手风琴式导航菜单切换设计的相关知识。
莫凡商城中可以通过手风琴式导航菜单切换图书分类菜单，如图 10.1 和图 10.2 所示。

 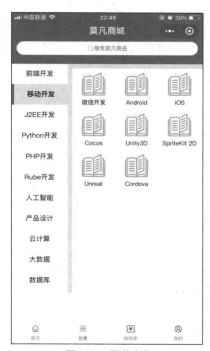

图 10.1　图书分类 1　　　　　　　图 10.2　图书分类 2

2. 任务实施

(1) 在 app.json 文件里添加图书分类页面路径 "pages/category/category"。

(2) 在 category.wxml 图书分类页面文件里进行图书分类布局设计，具体代码如下。

```
<view class="content">
  <view class="search">
    <view class="searchInput" bindtap="searchInput">
      <image src="/pages/images/tubiao/fangdajing-1.jpg" style="width:15px;height:19px;"></image>
      <text class="searchContent">搜索莫凡商品</text>
    </view>
  </view>
  <view class="category">
    <view class="left">
      <block wx:for="{{category}}">
        <view class="{{flag==index?'select':'normal'}}" id="{{index}}" bindtap="switchNav">{{item.firstTypeName}}</view>
      </block>
    </view>
    <view class="space"></view>
    <view class="right">
      <view class="hr"></view>
      <view class="rightContent">
        <swiper current="{{currentTab}}" style="height:500px;">
          <block wx:for="{{category}}">
            <swiper-item>
              <view class="items">
                <block wx:for="{{item.children}}" wx:for-item="it">
                  <view class="item" bindtap="more" data-firstid="{{item.firstId}}" data-secondid="{{it.secondId}}">
                    <view wx:if="{{it.secondTypeIcon}}">
                      <image src="{{it.secondTypeIcon}}"></image>
                    </view>
                    <view wx:else>
                      <image src="/pages/images/category/default.jpg"></image>
                    </view>
                    <view class="name">{{it.secondTypeName}}</view>
                  </view>
                </block>
              </view>
            </swiper-item>
          </block>
        </swiper>
      </view>
    </view>
  </view>
</view>
```

(3) 在 category.wxss 样式文件里进行图书分类样式渲染，具体代码如下。

```
.content {
  width: 100%;
  font-family: "Microsoft YaHei";
}
.search{
    width: 100%;
    background-color: #009966;
    height: 50px;
    line-height: 50px;
}
.searchInput{
    width: 95%;
    background-color: #ffffff;
    height: 30px;
    line-height: 30px;
    border-radius: 15px;
    display: flex;
    justify-content:center;
    align-items:center;
    margin: 0 auto;
```

```css
}
.searchContent{
    font-size:12px;
    color: #777777;
}
.category {
  display: flex;
  flex-direction: row;
}
.left {
  width: 30%;
  font-size: 15px;
}

.left view {
  text-align: center;
  height: 45px;
  line-height: 45px;
}
.select{
    background-color: #dddddd;
    border-left:5px solid #009966;
    font-weight: bold;
}
.normal{
    background-color: #ffffff;
    border-bottom: 1px solid #f2f2f2;
}
.space{
  width: 10px;
   background-color: #dddddd;
}
.right{
    width:70%;
}
.hr{
    height: 10px;
    background-color: #dddddd;
}
.line{
    height: 1px;
    width: 100%;
    background-color: #dddddd;
    opacity: 0.2;
}
.title{
  padding: 10px;
  font-size: 13px;
}
.items{
    display: flex;
    flex-wrap: wrap;
    justify-content: space-left;
}
.item{
  width: 33%;
  height: 80px;
  text-align: center;
  padding-top:10px;
}
.item image{
  width: 50px;
  height: 50px;
}
.name{
  font-size: 13px;
}
```

（4）在 category.js 业务逻辑处理文件里进行图书分类逻辑处理，具体代码如下。

```javascript
var app = getApp();
var host = app.globalData.host;
Page({
  data: {
    flag: 0,
    currentTab: 0,
    category: []
  },
  onLoad: function (options) {
    var page = this;
    this.loadCategory();
  },
  loadCategory:function(){
    var page = this;
    wx.request({
      url: host + '/api/category/getCategoryList',
      method: 'GET',
      data: { },
      header: {
        'Content-Type': 'application/json'
      },
      success: function (res) {
        console.log(res);
        var code = res.data.code;
        var category = res.data.data;
        if (code = '0000') {
          page.setData({ category: category });
        }
      }
    })
  },
  switchNav: function (e) {
    var page = this;
    var id = e.target.id;
    if (this.data.currentTab == id) {
      return false;
    } else {
      page.setData({ currentTab: id });
    }
    page.setData({ flag: id });
  },
  more:function(e){
    console.log(e);
    var firstId = e.currentTarget.dataset.firstid;
    var secondId = e.currentTarget.dataset.secondid;
    wx.navigateTo({
      url: '../goodsList/goodsList?firstId=' + firstId + "&secondId=" + secondId,
    })
  },
  searchInput: function (e) {
    wx.navigateTo({
      url: '../search/search',
    })
  }
})
```

10.6　项目实战：任务 23——实现图书分类结果列表功能

1. 任务目标

通过实现图书分类结果列表功能，来巩固图书列表动态获取的相关知识。

莫凡商城图书分类结果列表用来显示图书分类二级导航菜单对应的结果列表，结果列表包括图书名称、作者、出版社、出版时间及价格，如图 10.3 所示。

图 10.3　图书列表

2. 任务实施

（1）在 app.json 文件里添加图书列表页面路径"pages/goodsList/goodsList"。
（2）在 goodsList.wxml 页面文件里进行图书列表布局设计，具体代码如下。

```
<view class="content">
  <view class="search">
    <view class="searchInput" bindtap="searchInput">
      <image src="/pages/images/tubiao/fangdajing-1.jpg" style="width:15px;height:19px;"></image>
      <text class="searchContent">搜索莫凡商品</text>
    </view>
  </view>
  <view class="hr"></view>
  <view class="list">
    <block wx:for="{{books}}">
      <view class="book" bindtap="seeDetail" id="{{item.id}}">
        <view class="pic">
          <image src="{{item.listPic}}" mode="aspectFit" style="width:115px;height:120px;"></image>
        </view>
        <view class="movie-info">
          <view class="base-info">
            <view class="name">{{item.goodsName}}</view>
            <view class="desc">作者:{{item.author}} 著</view>
            <view class="desc">出版社:{{item.bookConcern}}</view>
            <view class="desc">出版时间:{{item.publishTime}}</view>
            <view class="people">
              <text class="price">￥{{item.goodsPrice}}</text>
              <text class="org">￥{{item.goodsCost}}</text>
            </view>
          </view>
        </view>
      </view>
      <view class="hr"></view>
    </block>
  </view>
</view>
```

慕课视频

项目实战：实现图书
分类结果列表功能

（3）在 goodsList.wxss 样式文件里进行图书列表样式渲染，具体代码如下。

```
.content{
    font-family: "Microsoft YaHei";
    width: 100%;
}
.search{
```

```
    width: 100%;
    background-color: #009966;
    height: 50px;
    line-height: 50px;
}
.searchInput{
    width: 95%;
    background-color: #ffffff;
    height: 30px;
    line-height: 30px;
    border-radius: 15px;
    display: flex;
    justify-content:center;
    align-items:center;
    margin: 0 auto;
}
.searchContent{
    font-size:12px;
    color: #777777;
}

.list{
    margin-top: 10px;
}
.book{
    display: flex;
    flex-direction: row;
    width: 100%;
}
.pic image{
    width:80px;
    height:100px;
    padding:10px;
}
.base-info{
    font-size: 12px;
    padding-top: 10px;
    line-height: 22px;
}
.name{
    font-size: 15px;
    font-weight: bold;
    color: #000000;
}
.people{
    color: #555555;
    margin-top: 5px;
    margin-bottom: 5px;
}
.price{
    font-size: 18px;
    font-weight: bold;
    color: #E53D30;
    margin-left:5px;
}
.org{
   text-decoration: line-through;
   margin-left: 10px;
   margin-right: 5px;
}
.desc{
    color: #333333;
}
.hr{
    height: 1px;
    width: 100%;
    background-color: #009966;
```

```
    opacity: 0.2;
}
```

（4）在 goodsList.js 业务逻辑处理文件里进行图书列表业务逻辑处理，具体代码如下。

```
var app = getApp();
var host = app.globalData.host;
Page({
  data: {
    books: null,
    host: host
  },
  onLoad: function(e) {
    var firstId = e.firstId;
    var secondId = e.secondId;
    this.getBookList(firstId, secondId);
  },
  getBookList: function(firstId, secondId) {
    var page = this;
    wx.request({
      url: host + '/api/goods/getGoodsList',
      method: 'GET',
      data: {
        firstId: firstId,
        secondId: secondId
      },
      header: {
        'Content-Type': 'application/json'
      },
      success: function(res) {
        var books = res.data.data;
        console.log(books);
        page.setData({
          books: books
        });
      }
    })
  },
  seeDetail: function(e) {
    var goodsId = e.currentTarget.id;
    wx.navigateTo({
      url: '../goodsDetail/goodsDetail?goodsId=' + goodsId,
    })
  },
  searchInput: function(e) {
    wx.navigateTo({
      url: '../search/search',
    })
  }
})
```

10.7 小结

本章讲解了设备应用相关 API 接口，包括获得系统信息、获取网络状态、加速度计、罗盘、拨打电话、扫码、剪贴板、蓝牙、屏幕亮度、震动、手机联系人；文件操作 API 接口，包括 wx.saveFile 保存文件到本地、wx.getSavedFileList 获取本地文件列表、wx.getSavedFileInfo 获取本地文件信息、wx.removeSavedFile 删除本地文件、wx.openDocument 打开文档、wx.getFileInfo 获取文件信息；窗口 API 接口，用来获取窗口大小参数；微信运动 API 接口，用来获取微信运动步数。